Using Formative Assessment to Drive Mathematics Instruction in Grades 3-5

Christine Oberdorf and Jennifer Taylor-Cox

Routledge
Taylor & Francis Group
LONDON AND NEW YORK

First published 2012 by Eye On Education

Published 2013 by Routledge
711 Third Avenue, New York, NY 10017, USA
2 Park Square, Milton Park, Abingdon, Oxon OX14 4RN

First issued in hardback 2015

Routledge is an imprint of the Taylor & Francis Group, an informa business

Cover Designer: Dave Strauss, 3FoldDesign

Library of Congress Cataloging-in-Publication Data

Oberdorf, Christine.
Using formative assessment to drive mathematics instruction in grades 3-5
/ by Christine Oberdorf and Jennifer Taylor-Cox.
p. cm.

1. Mathematics—Study and teaching (Elementary) 2. Educational tests and measurements.
3. Mathematical ability—Testing. I. Taylor-Cox, Jennifer.
II. Title.
QA11.2.O238 2011
372.7—dc23

2011012935

ISBN-13: 978-1-596-67190-4 (pbk)
ISBN-13: 978-1-138-16981-4 (hbk)

About the Authors

Ms. Christine Oberdorf is a dedicated and experienced educator. She serves as a Staff Development Teacher for Montgomery County Public Schools in Maryland in addition to working as a consultant in mathematics education. She is a National Board Certified Teacher and currently serves on the Board of Directors of the Maryland Council of Teachers of Mathematics. Christine appreciates opportunities to enable educators as they empower young mathematicians. She lives in Arnold, Maryland with her husband Arthur and her children Robert and Aley.

Dr. Jennifer Taylor-Cox is an enthusiastic, captivating presenter and well-known educator representing Taylor-Cox Instruction, LLC. Jennifer serves as a consultant in mathematics education and classroom discipline, providing professional development opportunities for elementary, middle school, and high school educators in numerous districts across the United States. Her keynote speeches, professional development workshops, strategic coaching models, classroom demonstration lessons, study groups, and parent workshops are always high-energy and insightful.

Jennifer earned her Ph.D. from the University of Maryland and was awarded the "Outstanding Doctoral Research Award" from the University of Maryland and the "Excellence in Teacher Education Award" from Towson University. She served as the president of the Maryland Council of Teachers of Mathematics. Jennifer is the author of numerous professional articles and educational books.

Jennifer knows how to make learning mathematics engaging and meaningful, motivating learners of all ages! She has a passion for using formative assessments to differentiate math instruction and solve classroom discipline issues. Jennifer lives and has her office in Severna Park, Maryland. She is the mother of three children. Jennifer truly understands how to connect research and practice in education. Her zeal for improving the quality of education is alive in her work with educators, students, and parents.

If you are interested in learning more about the professional development opportunities Dr. Taylor-Cox offers, please contact her at www.Taylor-CoxInstruction.com or 410.729.5599.

Taylor-Cox INSTRUCTION
Connecting Research and Practice in Education

Acknowledgements

Appreciation is expressed to…

Karren Schultz-Ferrell
Elementary Integrated Curriculum Instructional Specialist,
Montgomery County Public Schools
You are an expert in early childhood education and your feedback is invaluable.
Thank you for your friendship and encouragement.

Verna Washington
Instructional Specialist, Montgomery County Public Schools
You are always so willing to offer support. Thank you for sharing your knowledge and time.

Bonnie Ennis
Math Coordinator, Wicomico County Public Schools
Your perspective and feedback are greatly appreciated.

Contents

CCSS denotes concepts emphasized in the Common Core State Standards at the third through fifth grade levels.

1

achievement how resources

Introduction

what

mathematics student

What Is Formative Assessment?

Formative assessment drives mathematics instruction and is a key component in *Response to Intervention*. It is the process in which evidence of students' understanding is used by teachers to adjust instructional practice (Popham, 2008). As practitioners, we routinely monitor student performance on specific outcomes and standards. Formative assessments are employed to measure student performance so as to provide a targeted instructional response. Monitoring student learning through formative assessments provides a gauge, pinpointing where students are on the pathway of acquiring new knowledge. Their performance on these assessments provides work samples to analyze. The samples enable us to see where students are in comparison to where they need to be to meet the standard. Only through this process, are we equipped to provide an effective and meaningful instructional response. Without formative assessment, lesson planning is focused solely on curriculum with little regard for students' explicit academic needs.

How Does Formative Assessment Impact Student Achievement?

Effective formative assessment occurs simultaneously with instruction for the purpose of improving students' knowledge and performance in mathematics. When formative assessment is implemented properly, students learn what is being taught to a substantially greater degree (Black & Wiliam, 1998). When we provide feedback to students as a result of formative assessment, it is the most powerful factor in enhancing student achievement (Hattie & Jaeger, 1998). The National Mathematics Advisory Panel (2008) recommends regular use of formative assessment so that instruction can be adapted based on student progress. "Teachers' regular use of formative assessment improves their students' learning, especially if teachers have additional guidance on using the assessment to design and individualize instruction" (2008, p. xxiii). This book provides such guidance for teachers through a three-phase format of assessment, analysis, and response as illustrated below.

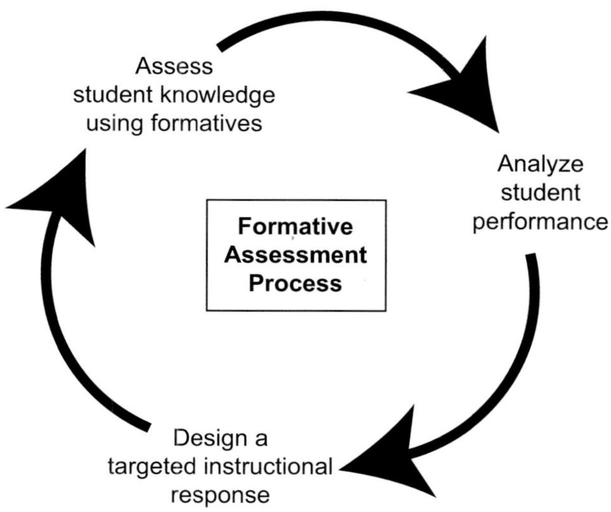

As practitioners, we experience tremendous pressure to "cover" the curriculum in a timely manner. Unfortunately, this sometimes translates to a practice of teaching *curriculum* rather than teaching *children*. Teaching and learning form a dynamic alliance that is reliant on the interactions between teachers and students. These interactions serve as feedback to teachers and inform next steps for instruction designed to advance learning. In order to efficiently and effectively teach children, we must understand what they already know in order to plan meaningful next steps.

We have heard teachers lament about this process and become overwhelmed at the prospect of providing differentiated instruction for individual students. Advocating an individualized instructional program is neither realistic nor appropriate for most classroom teachers. All students are entitled to instruction designed to meet their identified needs, but this does not have to translate into a one-on-one instructional setting. Students can be grouped according to similar instructional needs. When analyzing student understanding of a math concept for an entire class, patterns and trends emerge and students' needs are often revealed in clusters. There may be times when we need to work with an individual student to reteach a concept or clear up a misconception; however, often there is a small group of students for whom the data show similar academic needs.

How Is This Book Organized?

Using Formative Assessment to Drive Mathematics Instruction in Grades 3–5 contains seven chapters. The first chapter identifies the purpose and intentions of this book by describing formative assessment and highlighting the impact of the process on student performance. Chapters 2 through 6 outline a process for the use of formative assessment to inform instruction. Each of these chapters addresses one of five content standards in mathematics: number and operations; algebraic thinking; geometry; measurement; and data analysis and probability. Within each content standard, key mathematics concepts are highlighted. These concepts were identified from such sources as the *Principles and Standards of School Mathematics* (NCTM, 2000), the *Curriculum Focal Points for Prekindergarten through Grade 8 Mathematics* (NCTM, 2006), and the *Common Core State Standards Initiative* (2010). The concepts directly addressing the Common Core State Standards are labeled as such in the table of contents with this symbol **CCSS**. Other concepts serve as foundations for later CCSS. Chapter 7 is a brief conclusion with final comments of the formative assessment process.

The formative assessment process in Chapters 2 through 6 is presented in a three-page format for each highlighted skill or concept. Each of the three pages is designed to answer the following questions regarding student performance and mathematics instruction:

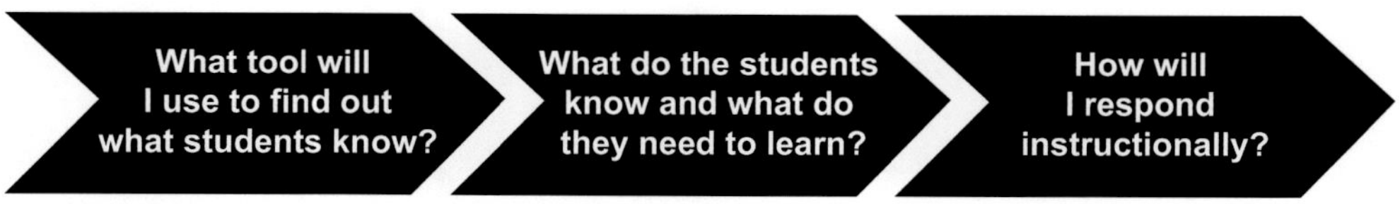

The formative assessments, student work samples, and suggested activities are provided for each mathematics concept to help teachers respond to these questions when planning instruction. Each is a deliberate step toward implementing effective mathematics instruction.

The **first** page in the three-page design illustrates a common sample of a *Traditional Formative Assessment* that one might find in a textbook or teacher resource (see Figure 1). Just below the assessment item is a *Limitations* note cautioning potential shortcomings of the traditional assessment. The *Traditional Formative Assessment* is then followed by a suggested *Enhanced Formative Assessment,* which is provided as an alternative to the traditional format. The *Enhanced Formative Assessment* is designed to elicit responses offering more insight into student understanding and remedy the *limitations* associated with the traditional assessment.

Figure 1

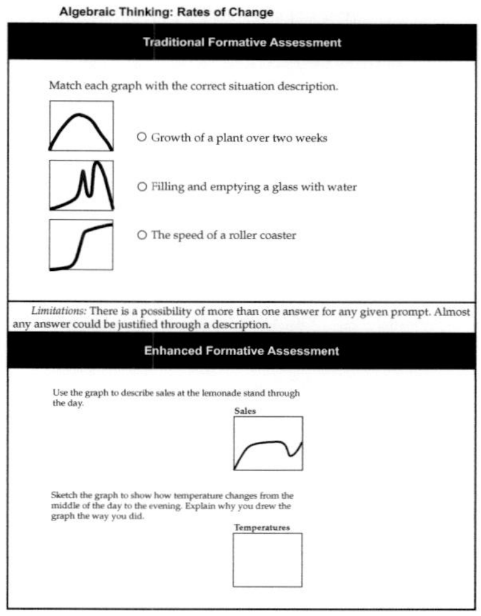

The **second** page in the three-page format includes *four student responses* to the enhanced formative assessment from the previous page (see Figure 2). The student responses are sequenced according to level of performance, beginning with a low level of understanding and ending with a high level of understanding for each concept. The first student work sample (Ulma) demonstrates minimal understanding. The second and third student samples (Vernon and Wayne) demonstrate moderate understanding, but with varying degrees of success. The final work sample (Yohanna) represents complete understanding of the assessed content. The shaded box just below each sample offers an analysis of the student work and guides the reader to the appropriate *Instructional Focus* provided on the next page.

Figure 2

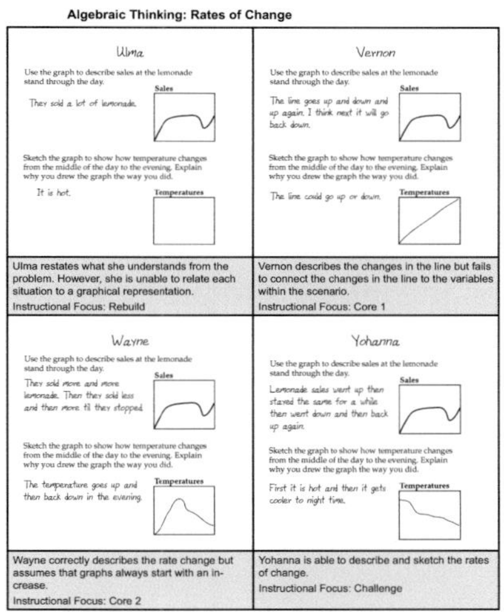

Although the instructional focus was developed with an individual student's response in mind, this is not to suggest that a teacher should develop or provide a different task for each student. It is not practical to expect teachers to implement individualized lessons. The student samples help to identify a level of understanding in which several students may fall. The task is intended to address the needs of students with similar understanding whom are working at a compatible level. The activities are designed to facilitate discourse among students in a small group setting.

The format of the **third** page includes four small group *Focus Activities* (see Figure 3). Each of the *Suggested Activities* is designed as a strategic response for students with instructional needs revealed in the work samples. The small group *Rebuild Focus* is designed for students demonstrating minimal understanding. The activity introduces the key concepts through an interactive task and provides an opportunity to acquire new knowledge. The small-group *Core 1 Focus* and small-group *Core 2 Focus* activities are geared for students with moderate levels of understanding. Although both *Core Focus* tasks target students with some understanding of the concept, each activity carries a different and specific focus based on the interpretation of needs from the student work samples. The *Core 1 Focus* activity is a slightly lower level of complexity than the *Core 2 Focus* activity. Lastly, the small-group *Challenge Focus* activity offers an enrichment opportunity for students demonstrating complete understanding of the math concept.

Each small-group *Focus Activity* is formatted with a defined goal, a list of needed materials, a description and directions for the task, and potential questions to pose while students are engaged in the activity. Response to Intervention calls on teachers to provide effective and meaningful instruction. This resource provides a means to that end by serving as a guide for identifying students' needs and responding accordingly.

How Can These Resources Be Utilized Effectively?

Figure 3

Algebraic Thinking: Rates of Change

	Rebuild Focus	Core 1 Focus	Core 2 Focus	Challenge Focus
	Talk It Out	What is Changing?	Ups and Downs	Constant or Variable
Goal	Connect rates of change to a graphic representation	Identify the attributes when viewing rates of change	Analyze graphs with decreasing rates of change	Collect data and display the rate of change
Materials	• Dry erase boards and markers	• Dry erase boards and markers • Graph paper	• Graph paper • Large index cards	• Graph paper
Suggested Activity Directions	Present students with familiar scenarios demonstrating rates of change. Ask students to describe the changes (e.g., number of students entering school in the morning, spending money at the mall, speed of a roller coaster ride). Provide time for students to sketch the changes as they consider the incremental changes. Focus on vocabulary such as increasing, decreasing, constant, and time. 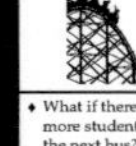	Ask students to name things that change over time. Provide examples such as: height, the amount of rainfall in a month, the growth of a flower, and money in a piggy bank. Ask students to discuss how these scenarios change over time. Demonstrate how students can sketch such changes as the attributes increase, decrease, or stay the same. When students are ready, they may transfer their sketches to graph paper. 	Share scenarios in which quantities might increase or decrease (e.g., temperature—morning to noon or noon to evening; allowance-spending or collecting; cookies in a cookie jar—baking or eating). Ask students to sketch a graph for each scenario. Students describe each graph on an index card. Shuffle graphs and cards and work as a group to match correct descriptions with correct graphs. 	Ask students to generate a list of data scenarios they could collect which would demonstrate a change over time (e.g., students ordering school lunch over a week's time, class attendance Monday to Friday, money spent at the school store, height of grass). Students simulate the scenario, collect, and display the data. Help students define the change as constant or variable. 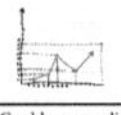
Questions to Assess	• What if there are more students on the next bus? • How is the amount of money changing as I buy more items? • Is the coaster moving at a constant speed?	• If the amount of money in the bank decreases, how is that decrease represented with the line? • What will the graph look like if your height stays the same for a long time?	• What key vocabulary helped you match the description to the correct graph? • How can you make your description specific enough so it only matches a single graph?	• Could you predict the rate of change for another time interval? • What attribute is changing over time? • How would you describe the degree of the change?

This book is designed to provide math educators with tools and resources targeting the varying needs of their students within five content standards of mathematics. When teaching a new concept or skill, students demonstrate a diverse range of competencies. A teacher introducing a lesson on fractions, for example, must simultaneously consider those students who are uncertain what a fraction represents, as well as those possessing a strong sense of fractions and are ready to tackle complex problems. The three-phase approach of this book assists teachers in identifying what students already know about the specific concept, facilitates meaningful student groupings according to current levels of understanding, and provides hands-on activities for instruction at four varying levels. It is not the intent of the authors that all students complete all four activities. Rather, the teacher may choose those activities most appropriate for groups of students.

Even though the activities are designed to be an appropriate and targeted response to the identified needs of a student, some students may need to engage in more than one of the provided activities. The sequence of task difficulty suggests that students beginning at the *Rebuild Focus* level may well benefit from working through the *Core 1* and *Core 2* tasks as well, in order to meet the standard of performance measured in the Enhanced Formative Assessment. Additionally, the four activities do not dictate that a teacher must always manage four groups. Rather, the intent is to provide four instructional activities from which to choose one that is best for meeting the instructional needs of the class. The sequence offered in the Table of Contents is not provided as a suggested order for teaching and should be adjusted to meet the specific scope and sequence of the district curriculum. Best wishes as you navigate the journey to student success by *Using Formative Assessment to Drive Mathematics Instruction in Grades 3–5.*

2

decimals fractions

Number & Operations

common equivalent division place value multiplication

CCSS Number & Operations: Place Value

Traditional Formative Assessment

What is the value of 7 in 470,853?

- O 700,000
- O 70,000
- O 7,000
- O 700

Limitations: A correct response could simply be a lucky guess. In addition, students choosing the same incorrect answer may have different instructional needs.

Enhanced Formative Assessment

Record two numbers greater than **470,853** using exactly the same digits. Explain how you know your numbers have a greater value.

CCSS Number & Operations: Place Value

Adam	Bianca
Record two numbers greater than **470,853** using exactly the same digits. Explain how you know your numbers have a greater value. 999,999 888,888 These are the biggest numbers you can make.	Record two numbers greater than **470,853** using exactly the same digits. Explain how you know your numbers have a greater value. 4,708,530 47,085,300 Longer numbers are bigger.
Adam's numbers are in fact larger than the number provided. However, by not using the digits provided, it is difficult to evaluate his understanding of place value. Instructional Focus: **Rebuild**	Bianca simply adds zeros to the original number in order to get a larger number, using too many digits rather than rearranging the digits to manipulate the place value. Instructional Focus: **Core 1**
Colton	**Diandre**
Record two numbers greater than **470,853** using exactly the same digits. Explain how you know your numbers have a greater value. 875,430 First, I put the digits in order from biggest to smallest so I would have the biggest number.	Record two numbers greater than **470,853** using exactly the same digits. Explain how you know your numbers have a greater value. 740,853 704,853 I just switched the first 2 numbers because 700,000 > 400,000. Then I switched the 10,000 and 1,000 digit for the second number for the same reason.
Colton uses the digits to create a single new number with a greater value. Instructional Focus: **Core 2**	Diandre correctly rearranges the digits to create two values larger than the original. Instructional Focus: **Challenge**

CCSS Number & Operations: Place Value

	Rebuild Focus	Core 1 Focus	Core 2 Focus	Challenge Focus
	Digit Switch	**Same Digits, Different Value**	**Number Sandwich**	**Target Difference**
Goal	Represent different values using the same digits	Sequence numbers according to value	Identify values between 2 numbers	Find the difference between 2 numbers
Materials	♦ Place value manipulatives such as Base Ten or Digi-Blocks ♦ Numeral cards 0–9 (see page 103)	♦ Numeral tiles (0–9) in a cup ♦ Sticky notes	♦ Six decahedron dice or 0–9 spinner (see page 103)	♦ Deck of playing cards (Ace through 9)
Suggested Activity Directions	Provide students with place value manipulatives. Shuffle numeral cards (0–9) and ask students to flip 4 cards over to make a 4-digit number. Ask students to represent the value using the place value manipulatives. Students switch the placement of 2 of the cards and discuss how the blocks must change to match the value of the new quantity. **1,234** **2,134**	Ask students to randomly draw 5 numeral tiles from a cup (containing tiles 0–9). Use the tiles to make a number and record on a sticky note. Rearrange the same tiles to create a different number. Record on a sticky note. Continue until 4 different numbers are made. Arrange the sticky notes in order from the least to the greatest value. 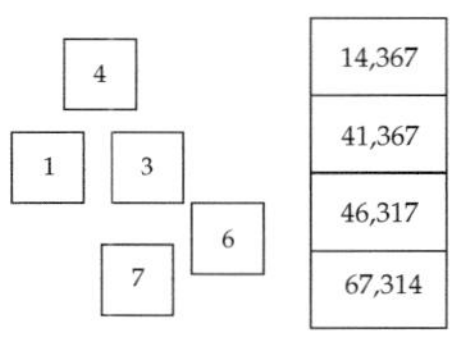	Ask students to roll 6 decahedron dice (0–9) and create a numeral. Repeat. Then ask students to identify 3 values that fall between the 2 values rolled. Prompt students to discuss how to generate potential numbers sandwiched between the numbers rolled.  **435,907** 436,000 501,654 543,210 **605,301**	Flip several playing cards over to generate a 6-digit number. Explore with students the difference between the original number and a new number created by switching 2 cards. Discuss how the difference changes when switching cards. Provide students with a target difference and ask them to decide which switch would result in the number closest to the target difference. **435,907** Which two digits should be switched to create a difference of about 200?
Questions to Assess	♦ How will the blocks change? Why? ♦ How did the switch in digits affect the value?	♦ What do you have to think about to put the values in order from least to greatest? ♦ What if the first 2 digits are the same?	♦ What 3 values can be sandwiched between the values rolled? ♦ How would you justify that they fit in the middle?	♦ How did you decide which 2 digits to switch? ♦ Which 2 digits would you switch to get the greatest difference? The least difference?

Traditional Formative Assessment

Solve each problem.

$$\begin{array}{r} 6 \\ \times\ 4 \\ \hline \end{array} \qquad 23 \times 8 = \qquad \begin{array}{r} 56 \\ \times\ 43 \\ \hline \end{array}$$

Limitations: Using a traditional algorithm to solve the problems does not guarantee a student possesses conceptual understanding of the operation.

Enhanced Formative Assessment

Show two ways to solve each problem.

$$\begin{array}{r} 6 \\ \times\ 4 \\ \hline \end{array} \qquad 23 \times 8 = \qquad \begin{array}{r} 56 \\ \times\ 43 \\ \hline \end{array}$$

CCSS Number & Operations: Multiplication of Whole Numbers

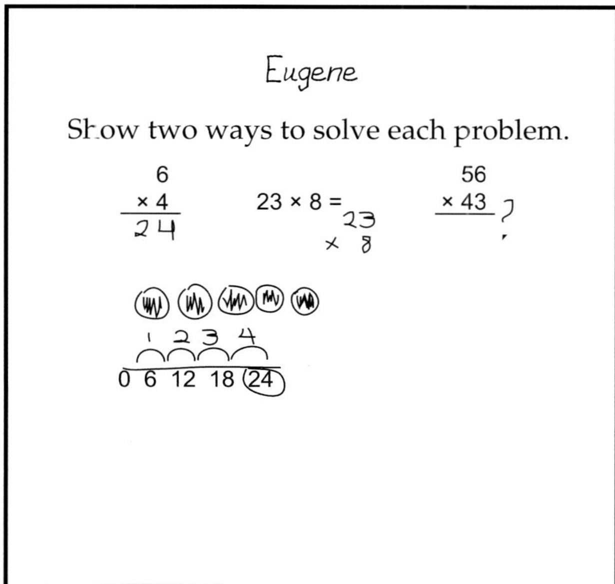

Eugene shows strategies for basic facts but is unable to apply strategies for larger quantities.
Instructional Focus: **Rebuild**

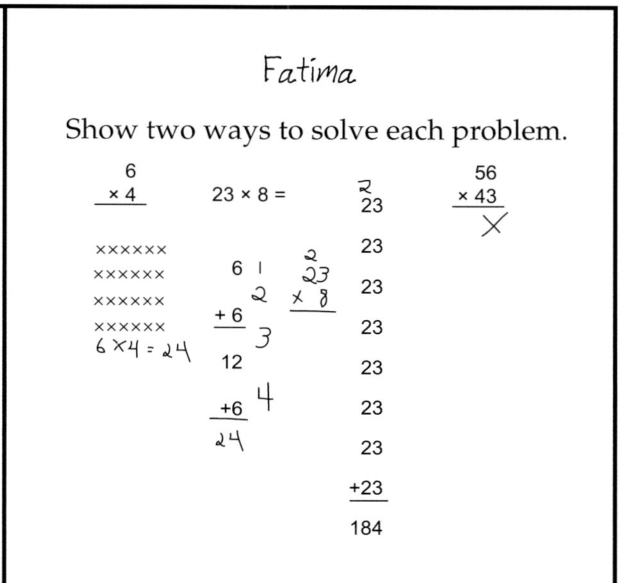

Fatima solves the basic fact and uses repeated addition to solve 23×8.
Instructional Focus: **Core 1**

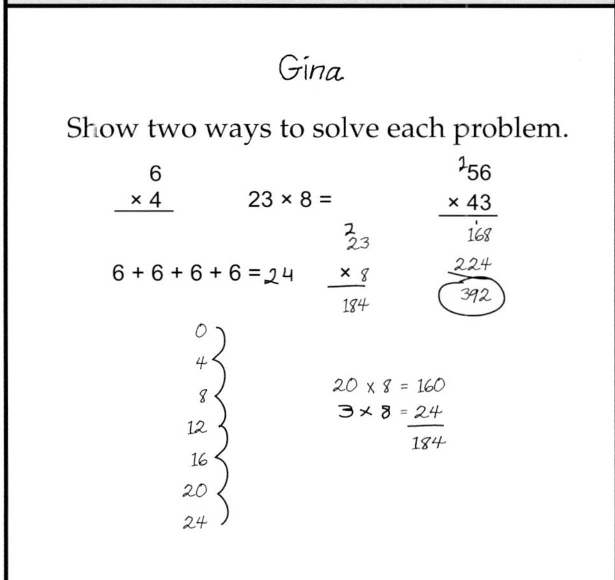

Gina correctly multiplies a 2-digit number by a single-digit number, but is unable to multiply two 2-digit numbers.
Instructional Focus: **Core 2**

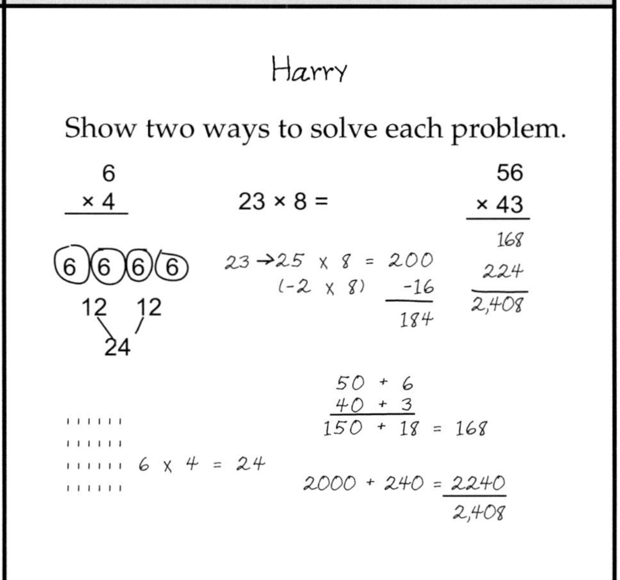

Harry uses accurate and efficient multiplication strategies.
Instructional Focus: **Challenge**

CCSS Number & Operations: Multiplication of Whole Numbers

	Rebuild Focus	Core 1 Focus	Core 2 Focus	Challenge Focus
	Break It Up!	**What's Close?**	**Area Model**	**High/Low**
Goal	Decompose 2-digit numbers to find a product	Apply derived facts to find products using 2-digit numbers	Use the area model to represent multiplication	Use place value to manipulate products
Materials	♦ 3 Numeral cubes (die)	♦ Numeral cube (die) ♦ Hundred Chart ♦ Chip (to drop on chart)	♦ Base 10 blocks	♦ Numeral cards (0–9) (see page 103)
Suggested Activity Directions	Ask students to roll the numeral cubes. Use the 3 digits to make a 2-digit by 1-digit multiplication problem. Ask students to break up the 2-digit number into expanded notation to show tens and ones. For example, 14 equals 10 + 4. $10 \times 6 = 60$ and $4 \times 6 = \underline{24}$ 84  **14** = 10 + 4 **×6** ×6 ×6 60 + 24 = 84	Ask students to close their eyes and drop the chip onto a hundred chart. This is the first factor. Roll the numeral cube to get a second factor. Students find the product using friendly facts to talk through the product. 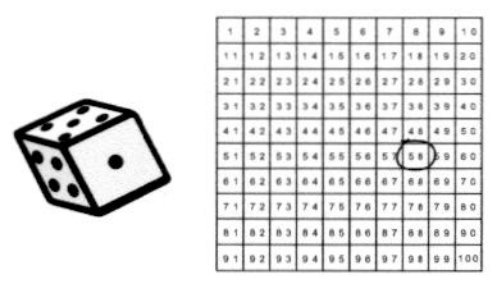 For example, 58 is close to 60. I know $60 \times 3 = 180$, so 58×3 is 6 less because I made the 58 into a 60 three times. The answer is 174.	Students use base 10 blocks to create an array for a multiplication problem. Set up the frame of the array using tens and ones blocks. Then fill in the array using hundreds, tens, and ones blocks. Students count to find the product. 22×13 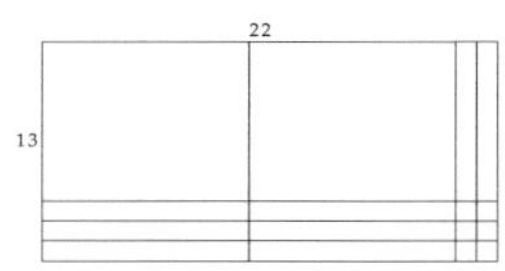 2 hundreds 8 tens 6 ones 286	Students randomly choose 4 cards. The object of the game is to arrange cards to set up a multiplication problem with the largest possible product. For example, the cards 1, 2, 3, and 4 could be arranged in any number of ways: 42 x 31 421 x 3 231 x 4 Find the arrangement that equals the largest possible product. Create the smallest possible product.
Questions to Assess	♦ How many tens? How many ones? ♦ What is the expanded form of the number? ♦ What happens when you multiply by a decade number?	♦ What friendly values could you use to estimate the product? ♦ What adjustments do you make after using the friendly numbers?	♦ How did you find the total of the partial products? ♦ How could you show your counting strategy in an equation?	♦ What strategy did you use to create your multiplication problem? ♦ Why is place value important in this game?

Traditional Formative Assessment

Choose the correct quotient.

854 ÷ 6 = ?

- O 14 r2
- O 142
- O 142 r2
- O 1,422

Limitations: An incorrect response fails to provide any diagnostic information to inform targeted instruction.

Enhanced Formative Assessment

The softball team sold 854 pizza kits and will put them in boxes to deliver. Each box holds 6 pizza kits. How many boxes are needed to deliver all of the pizza kits? Show your work.

CCSS Number & Operations: Division of Whole Numbers

Isaac

The softball team sold 854 pizza kits and will put them in boxes to deliver. Each box holds 6 kits. How many boxes are needed to deliver the kits?

$$\begin{array}{r} {}^{3\,2}854 \\ \times\ 6 \\ \hline 5{,}124 \end{array}\quad \text{boxes}$$

Isaac interprets the problem incorrectly and multiplies rather than divides to reach a solution.

Instructional Focus: **Rebuild**

Jonah

The softball team sold 854 pizza kits and will put them in boxes to deliver. Each box holds 6 kits. How many boxes are needed to deliver the kits?

$$\begin{array}{r} 14 \text{ r } 1 \\ 6\,\overline{)\,854} \\ -\ 6 \\ \hline 25 \\ -\ 24 \\ \hline 1 \end{array}$$

Jonah incorrectly divides using the traditional algorithm.

Instructional Focus: **Core 1**

Kenaja

The softball team sold 854 pizza kits and will put them in boxes to deliver. Each box holds 6 kits. How many boxes are needed to deliver the kits?

$$\begin{array}{r} 2 \\ 40 \\ 100 \\ 6\,\overline{)\,854} \\ 600 \\ \hline 254 \\ 240 \\ \hline 14 \\ 12 \\ \hline 2 \end{array}\quad 142 \text{ r2 boxes}$$

Kenaja correctly divides using partial quotients to find the quotient but does not interpret the remainder correctly in the answer.

Instructional Focus: **Core 2**

Lun

The softball team sold 854 pizza kits and will put them in boxes to deliver. Each box holds 6 kits. How many boxes are needed to deliver the kits?

$$\begin{array}{r} 142 \text{ r2} \\ 6\,\overline{)\,854} \\ 6 \\ \hline 25 \\ 24 \\ \hline 14 \\ 12 \\ \hline 2 \end{array}$$

$\frac{2}{6} = \frac{1}{3}$

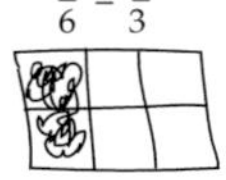

They need 143 boxes. All of the boxes will be filled but 1. It will have 2 kits and be 1/3 full.

Lun divides correctly and interprets the remainder to explain how many kits will be in the final box.

Instructional Focus: **Challenge**

CCSS Number & Operations: Division of Whole Numbers

	Rebuild Focus	Core 1 Focus	Core 2 Focus	Challenge Focus
	Operation Cards	**Look At It This Way…**	**Remainders Matter!**	**Picture This!**
Goal	Interpret a problem in order to choose the correct operation	Apply alternate algorithms to estimate and solve division problems	Interpret the remainder in a division problem	Divide a fraction by a whole number
Materials	♦ Operation Cards (see page 104) ♦ Counters	♦ Dry erase board and marker ♦ Place value manipulatives such as Base Ten or Digi-Blocks	♦ Blank index cards ♦ Chart paper	♦ Fraction pieces ♦ Grid paper (see page 106)
Suggested Activity Directions	Make a copy of the Operation Cards. Cut the cards apart and shuffle. Students take turns choosing a card and reading the problem to the group. They discuss whether they would use multiplication, division, or both to solve the problem. Once the group has come to consensus, students work together to solve the problem. Students may also write their own Operation cards to add to the card set. 	Students work in a small group to explore various representations and strategies to solve a single division problem such as 467 ÷ 3. Strategies may include: *Estimation* The quotient is more than 100 but less than 200 because 3 × 100 = 300 and 3 × 200 = 600. *Partial Quotients* 3 ⟌ 467 − 300 → 100 167 − 150 → 50 17 − 15 → + 5 2 → 155 r2	Ask students to write a division problem. Students create a real-life situation to match the division problem. Record each word problem on an index card. Students solve each problem and explain the meaning of the remainder. For example: The baseball team is going for ice cream after the big win. Each car holds 4 players. If there are 15 players in all, how many cars are needed? 	Students explore the fraction pieces and name the fraction represented by each. Present scenarios for students to divide a fraction by a whole number. For example: Sam wanted to share the remaining half of the pizza with 2 friends. How will the 3 people share half of the pizza? Students may use fraction pieces or draw a picture to represent the solution. 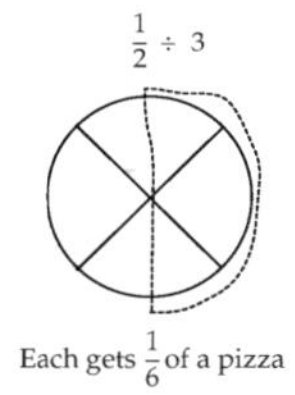
Questions to Assess	♦ Is this a multiplication or division situation? How do you know? ♦ How could we use manipulatives to act out the problem?	♦ Which mental math facts do you know that can help you estimate a quotient for this problem? ♦ How could you check your answer?	♦ Does the remainder change the answer? How do you know? ♦ Can the remainder be represented as a fraction?	♦ What picture can you draw to solve this problem? ♦ Does your answer make sense? How do you know?

Traditional Formative Assessment

What is ¼ of 12?

- O 1
- O 3
- O 4
- O 12

Limitations: The multiple choice format fails to require students to show their thinking. Student representations provide valuable insight into student understanding.

Enhanced Formative Assessment

Aley was drawing a picture of her candy pieces. She already drew ¼ of her candies. Draw the rest of Aley's candy pieces.

CCSS Number & Operations: Common Fractions

Michael

Aley was drawing a picture of her candy pieces. She already drew $\frac{1}{4}$ of her candies. Draw the rest of Aley's candy pieces.

Michael uses the 4 from the denominator and draws 4 candies.

Instructional Focus: **Rebuild**

Nikki

Aley was drawing a picture of her candy pieces. She already drew $\frac{1}{4}$ of her candies. Draw the rest of Aley's candy pieces.

Nikki incorrectly attempts to apply her understanding of fractional parts of a region to find the fractional parts of a set.

Instructional Focus: **Core 1**

Olamide

Aley was drawing a picture of her candy pieces. She already drew $\frac{1}{4}$ of her candies. Draw the rest of Aley's candy pieces.

Olamide shows one-fourth of 4 rather than one-fourth of 12.

Instructional Focus: **Core 2**

Pahana

Aley was drawing a picture of her candy pieces. She already drew $\frac{1}{4}$ of her candies. Draw the rest of Aley's candy pieces.

Pahana accurately represents one-fourth of 12.

Instructional Focus: **Challenge**

CCSS Number & Operations: Common Fractions

	Rebuild Focus	Core 1 Focus	Core 2 Focus	Challenge Focus
	Grab & Sort	**Sets Within Regions**	**Pet Store**	**Enough to Go Around**
Goal	Name the fractional parts of a set	Use fractional parts of a region as a work mat for fractional parts of a set	Explore fractional parts of a set with larger quantities	Solve more complex problems with fractions
Materials	♦ Different color cubes ♦ Paper or math journal	♦ Geoboard ♦ Geobands ♦ Connecting cubes	♦ Pet Store Cards (see page 105) ♦ Scissors	♦ Blank paper
Suggested Activity Directions	Students reach into a container of cubes and grab the largest possible handful. They then record the fractional part of each color in the set. 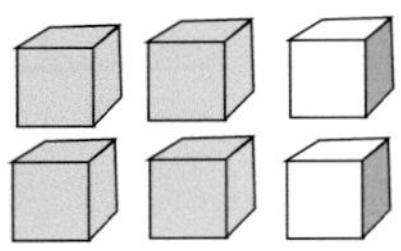 $\frac{4}{6}$ *are red* $\frac{2}{6}$ *are white* Enourage students to compare the sets. $\frac{2}{6}$ *is less than* $\frac{4}{6}$	Students use cubes to form fractional parts of sets. The geoboard serves as the work mat and is divided it into fractional parts (according to the prompts). Students distribute cubes among the regions to create equal-size sets. 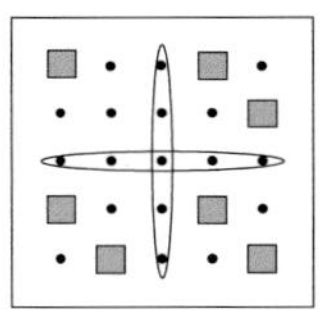 8 cubes into fourths 12 cubes into fourths 18 cubes into thirds 18 cubes onto sixths 24 cubes into sixths	Distribute a copy of the Pet Store Cards to pairs of students. Instruct students to cut out the fraction statements at the bottom of the page. Partners work together to match each statement with the correct card.  3/4 of the pets have a tail Provide time for students draw their own Pet Store card with fraction statements.	Ask students to draw pictures to solve problems in which they are dividing various numbers of regions into a specific number of groups. Allow students to share their strategies among the group to discuss and compare.  How could 10 pies be shared equally among 4 families?  How could 4 sandwiches be shared fairly by 6 children?  How could 5 friends share $3.00?
Questions to Assess	♦ How do you know what the numerator will be? ♦ How do you know what the denominator will be?	♦ How does the geoboard help to find the fractional parts of the set? ♦ How do you know how to divide up the geoboard?	♦ What is the total number of pets on the card? ♦ What fraction represents each type of pet on each card?	♦ How can you be certain that each person received a fair share? ♦ What if another person needed to be included in each problem?

CCSS Number & Operations: Equivalent Fractions

Traditional Formative Assessment

Choose the set that includes only equivalent fractions.

A. $\frac{1}{2}$ $\frac{1}{3}$ $\frac{1}{4}$

B. $\frac{1}{4}$ $\frac{2}{4}$ $\frac{3}{4}$

C. $\frac{1}{3}$ $\frac{2}{6}$ $\frac{3}{9}$

D. $\frac{1}{2}$ $\frac{2}{3}$ $\frac{3}{4}$

Limitations: As with all multiple choice questions, correct and incorrect answers provide limited assumptions regarding student learning. Additionally, the format of this problem makes it difficult to interpret with so many fractions so close together.

Enhanced Formative Assessment

List two fractions equivalent to ¼.
Draw a picture to justify your answer.

CCSS Number & Operations: Equivalent Fractions

Quinn

List two fractions equivalent to $\frac{1}{4}$.
Draw a picture to justify your answer.

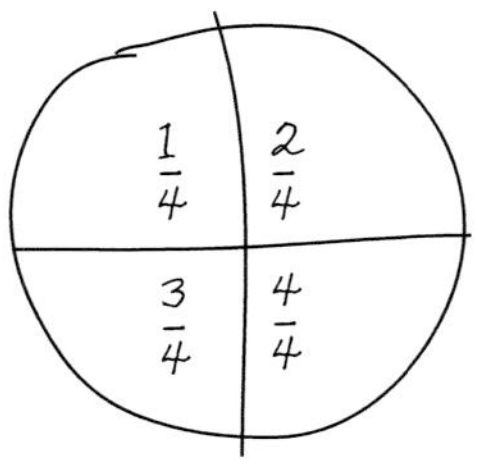

Quinn demonstrates his ability to count by fourths but does not show any concept of equivalence.

Instructional Focus: **Rebuild**

Ryan

List two fractions equivalent to $\frac{1}{4}$.
Draw a picture to justify your answer.

$$\overset{+1 \quad +1}{\frac{1}{4} = \frac{2}{5} = \frac{3}{6}}_{+1 \quad +1}$$

Ryan attempts to repeat a pattern to find equivalent forms for ¼.

Instructional Focus: **Core 1**

Sabastine

List two fractions equivalent to $\frac{1}{4}$.
Draw a picture to justify your answer.

$\frac{4}{16}$

I drew a box and colored one-fourth. Then I divided again and again. I just counted all the pieces and the pieces shaded.

Sabastine correctly represents equivalent forms for ¼ but only names one fraction.

Instructional Focus: **Core 2**

Tyra

List two fractions equivalent to $\frac{1}{4}$.
Draw a picture to justify your answer.

$$\frac{1 \times 2}{4 \times 2} = \frac{2}{8}$$

$$\frac{1 \times 3}{4 \times 3} = \frac{3}{12}$$

Tyra finds two fractions equivalent to ¼ by multiplying both the numerator and denominator by the same factor.

Instructional Focus: **Challenge**

CCSS Number & Operations: Equivalent Fractions

	Rebuild Focus	Core 1 Focus	Core 2 Focus	Challenge Focus
	Fraction Scramble	**Number Lines**	**Grid Paper Galore**	**Multiplication Chart**
Goal	Identify equivalent fractions using manipulatives	Compare fractions to confirm equivalent values	Represent equivalent forms of a fraction	Justify why two fractions are equivalent
Materials	♦ Fraction cards ♦ Fraction pieces (halves, thirds, and fourths)	♦ Grid paper (see page 106)	♦ Grid paper (see page 106)	♦ Multiplication chart (see page 107)
Suggested Activity Directions	Record each of the fractions listed below on an index card. Shuffle the cards. Students sort cards into piles of equivalent fractions using fractions pieces to confirm each sort. Students should explain their equivalent fractions as the cards are sorted. $\frac{1}{2}$, $\frac{2}{4}$, $\frac{3}{6}$, $\frac{1}{3}$, $\frac{2}{6}$, $\frac{3}{9}$ $\frac{1}{4}$, $\frac{2}{8}$, $\frac{3}{12}$ 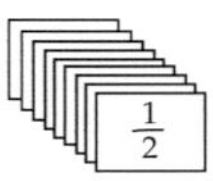 	Students draw 3 number lines of equal length. On the first, mark and label 1/4. On the second, mark and label 2/5. On the third, mark and label 3/6. Prompt students to recall where 1/2 is located. As students estimate the distance for each value, they see that the three values are different distances from 0 and 1 and are not equal. Repeat using 1/2, 2/4, and 3/6. $\frac{1}{2}$ $\frac{2}{4}$ $\frac{3}{6}$ 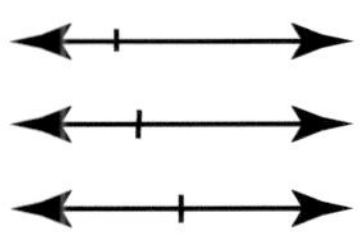	Draw 3 rectangles that can be divided into fourths (grid paper makes this easier and more accurate for students). Illustrate different ways to divide the rectangle into fourths (1/4 , 2/8, 4/16). This strategy prevents students from confusing the fractional parts and the whole from one model to the next. $\frac{1}{2}$ $\frac{2}{8}$ $\frac{4}{16}$ 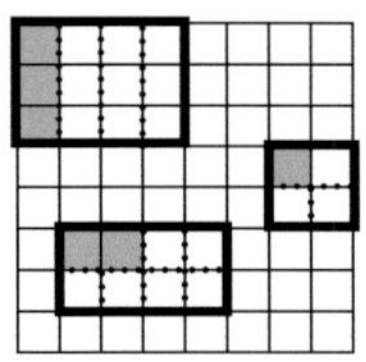	Use a multiplication chart to illustrate equivalent fractions by reading the vertical numerals as fractions and recognizing the equivalent fractions across the table. This provides a visual for students to see that both the numerator and denominator are being multiplied by the same factor. In fraction form, this has a value of 1 and demonstrates the identity property for multiplication. (multiplication chart below)
Questions to Assess	♦ Using the fraction pieces, how many equivalent fractions can you find for ½? For ¼? ♦ How do you know the fractions are equal?	♦ Is it closer to 0 or 1 whole? ♦ Is the fraction larger or smaller than ½?	♦ What other denominators can be seen in your representation? ♦ What may the representation look like for thirds?	♦ What patterns do you notice? ♦ Why does multiplying the numerator and denominator by the same factor result in an equivalent fraction?

X	1	2	3	4	5	6	7	8	9
1	1	2	3	4	5	6	7	8	9
2	2	4	6	8	10	12	14	16	18
3	3	6	9	12	15	18	21	24	27
4	4	8	12	16	20	24	28	32	36
5	5	10	15	20	25	30	35	40	45
6	6	12	18	24	30	36	42	48	54
7	7	14	21	28	35	42	49	56	63
8	8	16	24	32	40	48	56	64	72
9	9	18	27	36	45	54	63	72	81

Traditional Formative Assessment

Solve:

$\frac{1}{4} + \frac{2}{4} =$ $\quad\quad$ $\frac{1}{3} + \frac{3}{5} =$ $\quad\quad$ $4\frac{1}{6} - 2\frac{2}{4} =$

Limitations: Without prompting, students may not show or explain the strategy they used to solve the problem. Additionally, computation skills do not always equate to conceptual understanding.

Enhanced Formative Assessment

How would you tell a friend to solve each problem?

$\frac{1}{4} + \frac{2}{4} =$

$\frac{1}{3} + \frac{3}{6} =$

$4\frac{1}{6} - 2\frac{2}{4} =$

Ulri

How would you tell a friend to solve each problem?

$\frac{1}{4} + \frac{2}{4} =$ add the tops and the botums

$\frac{1}{3} + \frac{3}{6} =$ X

$4\frac{1}{6} - 2\frac{2}{4} =$ X

Ulri tries to apply addition of whole numbers to the addition problem with fractions.

Instructional Focus: **Rebuild**

Vanessa

How would you tell a friend to solve each problem?

$\frac{1}{4} + \frac{2}{4} =$ add the numerators and keep the 4 on the bottom.

$\frac{1}{3} + \frac{3}{6} =$ you cannot do this becus bottom numbers are not same

$4\frac{1}{6} - 2\frac{2}{4} =$ don't know

Vanessa is able to add fractions with like denominators but lacks strategies when computing fractions with unlike denominators.

Instructional Focus: **Core 1**

William

How would you tell a friend to solve each problem?

$\frac{1}{4} + \frac{2}{4} =$ count from $\frac{1}{4}$ to 2 more forths $\frac{1}{4}$ $\frac{2}{4}$ $\frac{3}{4}$

$\frac{1}{3} + \frac{3}{6} =$ $\frac{1}{3}$ is the same as $\frac{2}{6}$ so add 2 to 3 for $\frac{5}{6}$

$4\frac{1}{6} - 2\frac{2}{4} =$ denomators don't work

William correctly adds fractions with like denominators and fractions in which one denominator is a factor of the second.

Instructional Focus: **Core 2**

Xavier

How would you tell a friend to solve each problem?

$\frac{1}{4} + \frac{2}{4} =$ add the numerator- the denomator stays the same because they are both forths.

$\frac{1}{3} + \frac{3}{6} =$ You have to find a common denomator. It is 6. Change the first numerator to 2. Add the numerators.

$4\frac{1}{6} - 2\frac{2}{4} = 4\frac{2}{12} - 2\frac{6}{12} = 3\frac{14}{12} - 2\frac{6}{12} =$

$1\frac{8}{12} = 1\frac{4}{6} = 1\frac{1}{3}$

Xavier accurately adds and subtracts fractions and mixed numbers with like and unlike denominators.

Instructional Focus: **Challenge**

CCSS Number & Operations: Addition & Subtraction of Fractions

	Rebuild Focus **Counting Up & Back**	Core 1 Focus **Making One Whole**	Core 2 Focus **Ruler Raceway**	Challenge Focus **Domino Search**
Goal	Count forward and backward by common fractional parts	Find equivalent forms of a fraction using manipulatives	Counting up and back with unlike denominators	Name simplest form of a fraction
Materials	♦ Pattern blocks ♦ Isometric grid paper (see page 108)	♦ Fraction pieces	♦ 12″ ruler ♦ Mini-car or chip ♦ Cube with fractions 1/2, 1/4, 1/8, 2/4, 3/6, 2/8	♦ Double 12 Dominoes
Suggested Activity Directions	Provide students time to name the fractional parts of a hexagon using triangles, rhombuses, and trapezoids. Students collect several blocks of a common shape and count the fraction pieces. Ask students to count forward and backward by the common fractions. Record the counting as addition and subtraction equations. 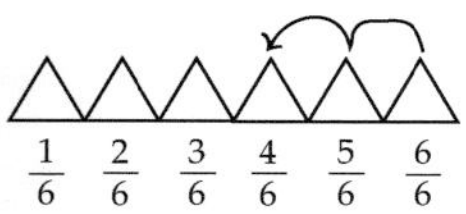 $\frac{6}{6} - \frac{2}{6} = \frac{4}{6}$	Students explore the fraction pieces and identify the value of each. Ask students to trace the shape representing one whole and use the pieces to fill in the whole two different ways (denominators). Challenge students to find several different ways to make one whole by mixing denominators. Students label the pieces used to make one whole. 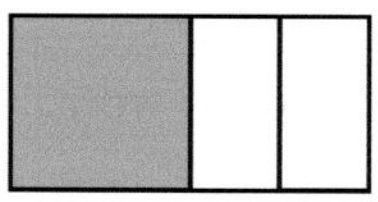 $\frac{1}{2} + \frac{2}{4} = 1$	A ruler serves as the raceway. Students place their car at the start (0). On turn, each student rolls the die and moves the designated distance (in inches). Each player marks and records the new location on the blank sheet of paper under the ruler. Players take turns racing to the 6-inch mark, recording the new location after each move. Students may also race back from 6 inches to explore subtraction.   ½ , 2, 2¼	Each domino is read as a fraction. Ask students to search for all of the dominos for which the numerator and denominator share a common factor. Once all of the dominos are found, students identify the common factor and name the simplified fraction. Ask students to match the domino with the domino showing the simplified fraction. 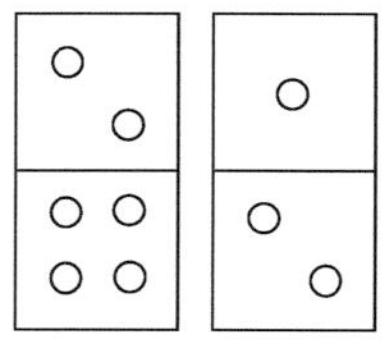
Questions to Assess	♦ Is counting forward a model of addition or subtraction? ♦ How would you record your counting as an addition or subtraction problem?	♦ Can you make ½ another way? ♦ Which denominators did you use to make one whole?	♦ How will you find your new location? ♦ When the denominators are different, how do you know how far to move?	♦ How do you know the fractions are equivalent? ♦ Is this fraction in simplest form? How do you know?

Traditional Formative Assessment

Circle the correct equations. Put an X on the incorrect equations.

$.7 + 4 = .74$	$1.3 + .7 = 2.0$	$.5 + .5 = .10$	$1.67 + .123 = .2810$
$1.5 - .7 = 8$	$.345 - .34 = .5$	$1.0 - .7 = .3$	$.456 - .01 = .446$

Limitations: Correct or incorrect responses could simply be caused by the layout of the problems rather than as a gauge of student understanding.

Enhanced Formative Assessment

Some of these problems are incorrect. Fix the errors.

$.7 + 4 = .74$ $\quad$ $1.3 + .7 = 2.0$ $\quad$ $1.5 - .7 = 8$

$$\begin{array}{r} 1.67 \\ +.123 \\ \hline 2.90 \end{array} \qquad \begin{array}{r} 345 \\ -.34 \\ \hline .005 \end{array} \qquad \begin{array}{r} 2.00 \\ -.65 \\ \hline 2.65 \end{array}$$

CCSS Number & Operations: Addition & Subtraction of Decimals

Yohanna Some of these problems are incorrect. Fix the errors. .7 + 4 = .74 ✓ 1.3 = .7 = 2.0 ✓ 1.5 - .7 = 8 ✓ 1.67 + .123 = 2.90 ✓ .345 - .34 = .005 (.345 - .34 = 311) 2.00 - .65 = 2.65 ✓	**Zareb** Some of these problems are incorrect. Fix the errors. .7 + 4 = .74 (4.7) 1.3 = .7 = 2.0 1.5 - .7 = 8 (.8) 1.67 + .123 = 2.90 ✓ .345 - .34 = .005 ✓ 2.00 - .65 = 2.65
Yohanna is unable to distinguish between correct and incorrect responses and erroneously subtracts because she incorrectly aligns the values. Instructional Focus: **Rebuild**	Zareb computes to the tenths place. He is likely able to solve these problems mentally. His difficulty lies in computing decimals to the hundredths and thousands place. Instructional Focus: **Core 1**
Aaron Some of these problems are incorrect. Fix the errors. .7 + 4 = .74 (4.7) 1.3 = .7 = 2.0 1.5 - .7 = 8 (.8) 1.67 + .123 = 2.90 .345 - .34 = .005 2.00 - .65 = ~~2.65~~ 1.35	**Becca** Some of these problems are incorrect. Fix the errors. .7 + 4 = ~~.74~~ 4.7 1.3 = .7 = 2.0 1.5 - .7 = .8 1.67 + .123 = ~~2.90~~ (1.67 + .123 = 1.793) .345 - .34 = .005 2.00 - .65 = ~~2.65~~ 1.35
Aaron correctly fixes all errors with the exception of the addition to the thousandths place when the decimals are not aligned properly. Instructional Focus: **Core 2**	Becca effectively computes to correct any errors. Instructional Focus: **Challenge**

CCSS **Number & Operations: Addition & Subtraction of Decimals**

	Rebuild Focus	Core 1 Focus	Core 2 Focus	Challenge Focus
	Racing for 5!	**Money, Money, Money**	**Hit the Target**	**Times What?**
Goal	Add and subtract tenths using models	Add and subtract hundredths using money	Add and subtract decimals to the thousandths place	Multiply a decimal by a whole number
Materials	♦ Place value manipulatives such as Base Ten or Digi-Blocks ♦ Number cube (1–6)	♦ Variety of coins and bills ♦ Ads from magazines and newspapers	♦ Decahedron die (0–9) ♦ Target Game Board (see page 109)	♦ Magazine and newspaper ads ♦ Variety of coins and bills ♦ Decahedron die (0–9)
Suggested Activity Directions	Distribute a mat to each student to help identify the work space. The object of the game is to make 5 wholes. Students roll the number cube and collect the number of tenths. They add the tenths to work mat and record the new total. Play continues until students make 5 wholes using tenths and trading for tenths and wholes. They may also race back from 5 to 0. I added | I now have .2 | .2 .5 | .7 .4 | 1.1	Distribute ads and play money to students. Assign tasks in which students add and subtract monetary amounts. Ask students to show their work. Sample tasks include: ♦ Choose five items and find the total price. ♦ Find the difference in price between any two items. ♦ How might you spend \$20 for a gift? ♦ How much more did you spend than your neighbor? 	Each student gets a game board. One player rolls the decahedron die. All students must use the rolled digit and place it in one of the blanks on the game board (each player may choose a different space). When the game board is completely filled, students receive a point for each equation or expression that is correct. Students compare their scores and discuss their strategies for placing each digit. 	Distribute ads and ask students to browse for items they might choose to buy. Once they choose an item, students roll the decahedron die to determine how many of the item they will purchase. Students may use the play money or develop their own algorithms to find each product. Students share their multiplication strategies within the group. 
Questions to Assess	♦ After adding the tenths, what is your new total? ♦ How many more tenths do you need to make the next whole? ♦ How far are you from 5 wholes?	♦ How much money have you spent? ♦ How much do you have left? ♦ How do you know where to place the decimal point when adding or subtracting money?	♦ Why did you choose that space for the digit? ♦ Is the equation / expressions correct? How do you know? ♦ What would you do differently if you played again?	♦ How could you calculate the product? ♦ About how high should your total be? ♦ Is your strategy accurate and efficient?

3

Algebraic Thinking

geometric
change
pattern
variables
equalities
numeric
reasoning
functions

CCSS Algebraic Thinking: Numeric Patterns

Traditional Formative Assessment

List the next number in each pattern.

15, 13, 11, 9, 7, ___

1, 2, 4, 8, 16, ___

Limitations: Requesting students to provide only the next single term gives limited insight into a students' thinking, which makes it difficult to plan targeted instruction.

Enhanced Formative Assessment

List the next three terms in each pattern.

15, 13, 11, 9, 7, ___, ___, ___, ...

1, 2, 4, 8, 16, ___, ___, ___, ...

Create a different numeric pattern.

___, ___, ___, ___, ___, ___, ...

CCSS Algebraic Thinking: Numeric Patterns

Allison	Byron
List the next three terms in each pattern. 15, 13, 11, 9, 7, <u>15</u>, <u>13</u>, <u>11</u>, ... 1, 2, 4, 8, 16, <u>1</u>, <u>2</u>, <u>4</u>, ... Create a different numeric pattern. <u>7</u>, <u>6</u>, <u>5</u>, <u>7</u>, <u>6</u>, <u>5</u>, ...	List the next three terms in each pattern. 15, 13, 11, 9, 7, <u>5</u>, <u>3</u>, <u>1</u>, ... 1, 2, 4, 8, 16, <u>24</u>, <u>32</u>, <u>40</u>, ... Create a different numeric pattern. <u>1</u>, <u>2</u>, <u>3</u>, <u>4</u>, <u>5</u>, <u>6</u>, ...
Allison used her knowledge of repeating patterns to continue the sequence. She saw the given terms as the core and continued the pattern by repeating the terms provided. Instructional Focus: **Rebuild**	Byron accurately continued the first patterns, increasing by 2. However, in the second sequence, he limited his analysis to the difference between the fourth and fifth terms and continued the sequence by adding 8 each time. Instructional Focus: **Core 1**
Chris	**Dante**
List the next three terms in each pattern. 15, 13, 11, 9, 7, <u>5</u>, <u>3</u>, <u>1</u>, ... 1, 2, 4, 8, 16, <u>32</u>, <u>64</u>, <u>128</u>, ... Create a different numeric pattern. <u>1</u>, <u>3</u>, <u>5</u>, <u>7</u>, <u>9</u>, <u>11</u>, ...	List the next three terms in each pattern. 15, 13, 11, 9, 7, <u>5</u>, <u>3</u>, <u>1</u>, ... 1, 2, 4, 8, 16, <u>32</u>, <u>64</u>, <u>128</u>, ... Create a different numeric pattern. <u>5</u>, <u>5</u>, <u>10</u>, <u>15</u>, <u>25</u>, <u>40</u>, ...
Chris successfully continued both growing patterns. She recognized the doubling nature of the second sequence. She also created a new growing sequence and is ready to learn about more complex numeric patterns. Instructional Focus: **Core 2**	Dante clearly continued the numeric patterns and created a complex pattern in which the 2 preceding numbers are added to form the next number. Instructional Focus: **Challenge**

CCSS Algebraic Thinking: Numeric Patterns

	Rebuild Focus	Core 1 Focus	Core 2 Focus	Challenge Focus
	Rollin' Patterns	**Dissecting Patterns**	**Close Your Eyes**	**Below 0 and <1**
Goal	Extend growing numeric patterns	Identify the progression from one term to the next in a growing pattern	Create complex numeric patterns	Extend numeric patterns, including negative numbers and/or fractions
Materials	♦ Dry erase boards and markers ♦ Die (0–9) ♦ Die (labeled + + + − − −)	♦ Chart paper ♦ Markers ♦ 1½″ × 2″ Sticky notes	♦ Dry erase boards and markers ♦ Index cards ♦ Hundred chart	♦ Number lines (−20 through +20) ♦ Fraction pieces ♦ Dry erase boards and markers
Suggested Activity Directions	Show students how to draw a series of blanks separated by commas. Ask a student to roll the numeral cube and the operation cube. Explain that these cubes determine how the pattern grows. Once the operation and value are determined, ask students to discuss what the first value might be. Ask students to continue the pattern. − 4 100, 96, 92, 88, 84	Write a numeric pattern on chart paper. Leave a large space between each number in the sequence. Examples: 2, 4, 6, 8, 10, … 21, 18, 15, 12, 9, … 1, 2, 4, 8, 16, 32, … Ask students to use mini sticky notes to label the difference between the terms and place the notes between the terms. +2 +2 +2 +2 2, 4, 6, 8, 10,	Explain to students what the rule will be for a growing pattern. Possibilities include: ♦ Add 2 ♦ Multiply by 3 ♦ Subtract 5 ♦ Divide by 2 ♦ Double and add 1 Ask a student to close his eyes and point to the hundred chart. The number he randomly chooses must be used in the sequence in any position. Allow students to create a sequence using the number and the rule. 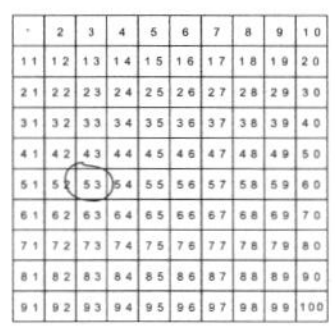	Read one of the following patterns as students record it on the dry erase board: 10, 7, 4, 1, … −7, −4, −1, 2, … $\frac{1}{2}$, 1, $1\frac{1}{2}$, 2, … 8, 4, 2, 1, … 81, 27, 9, 3, 1, … Provide opportunity for students to use the number line and fraction pieces to extend each pattern. Students can also create their own patterns that include fractions or negative numbers.
Questions to Assess	♦ How might the operation determine the number you choose as the first in the sequence? ♦ How could you check your work?	♦ Are the values increasing or decreasing? ♦ What do you notice about the differences you wrote on the notes?	♦ Where in the sequence will you use the value? Why? ♦ Will the rule cause an increase or decrease as the sequence progresses?	♦ What is the difference between the two terms? ♦ What is the next number in the sequence? How do you know?

Traditional Formative Assessment

How many square tiles are needed to show the next figure in the pattern?

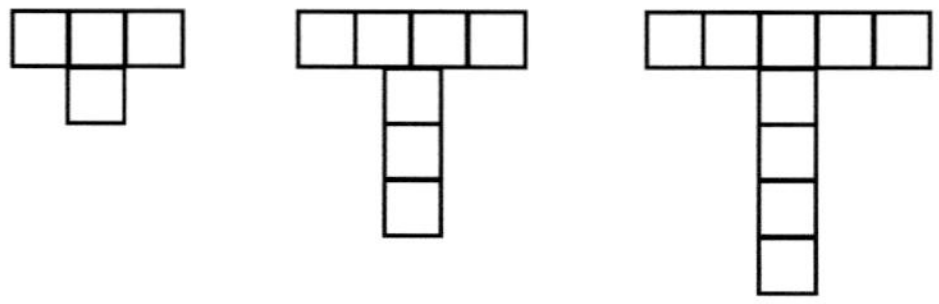

- ○ 4 tiles
- ○ 10 tiles
- ○ 13 tiles
- ○ 15 tiles

Limitations: As with other multiple choice questions, the student may not know the answer but simply have a lucky guess. Students also need opportunities to represent a correct answer.

Enhanced Formative Assessment

Continue repeating the pattern.

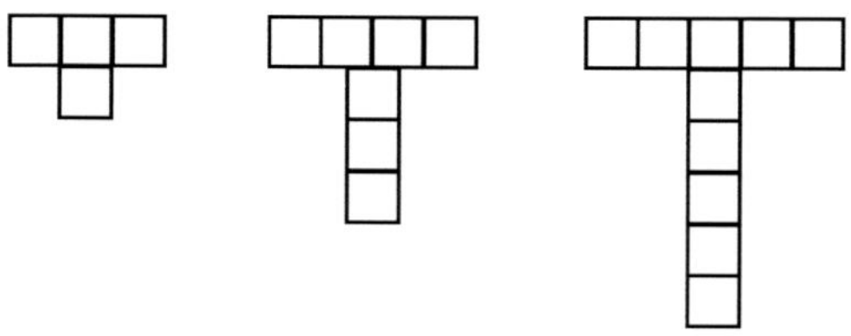

Describe the pattern using words or numbers.

Elaine

Continue repeating the pattern.

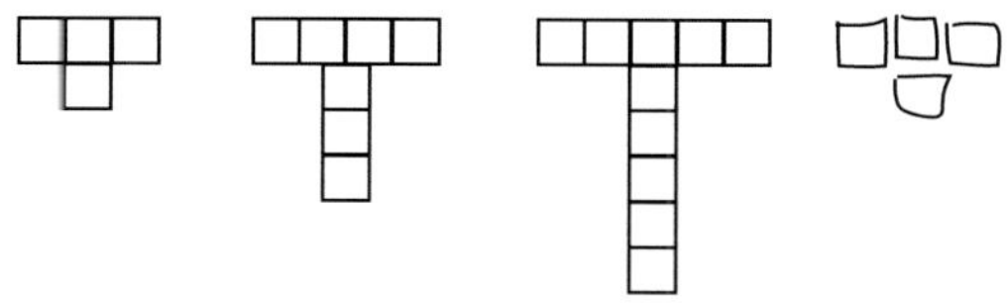

Describe the pattern using words or numbers.

I startd over.

Elaine used the pattern as a core and began repeating the designs rather than extending the pattern.

Instructional Focus: **Rebuild**

Franklin

Continue repeating the pattern.

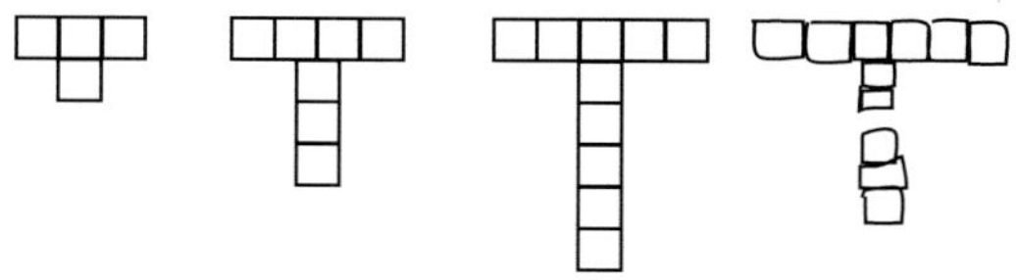

Describe the pattern using words or numbers.

I added 1 to the top.

Franklin recognized that the rows increased by 1 each time and accurately drew 6 squares. However, Franklin did not identify the increasing nature of the vertical column.

Instructional Focus: **Core 1**

George

Continue repeating the pattern.

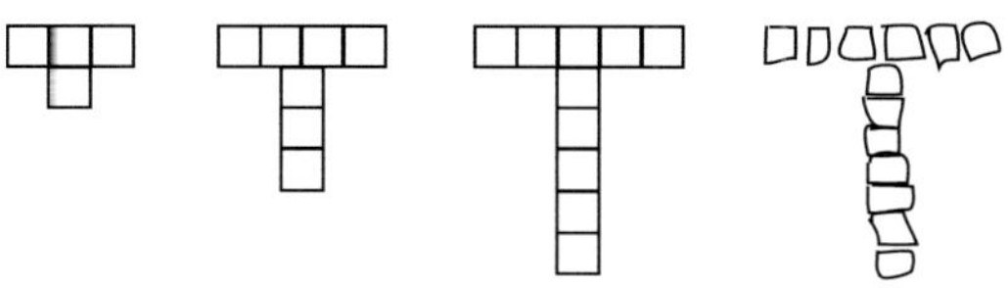

Describe the pattern using words or numbers.

I added more sqares acrooss and down.

George correctly continued the pattern by increasing the row by 1 and the column by 2 squares. He is ready to learn more about ways of describing growing patterns.

Instructional Focus: **Core 2**

Helena

Continue repeating the pattern.

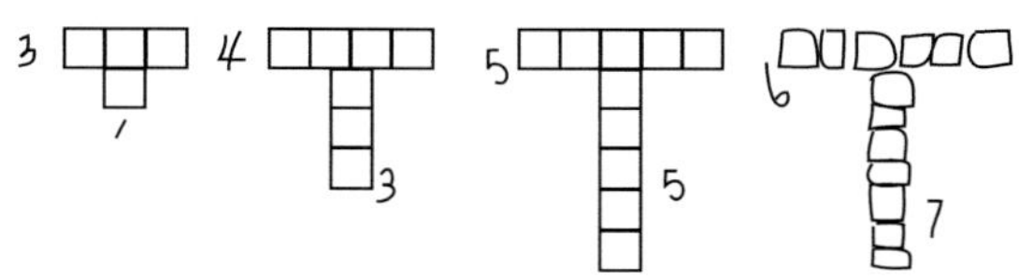

Describe the pattern using words or numbers.

The horizontal row gets 1 bigger and the vertical gets 2 bigger each time.

Helena continued the pattern and described the progression using words and by labeling the figures with numbers to represent each increase.

Instructional Focus: **Challenge**

CCSS Algebraic Thinking: Shape Patterns

	Rebuild Focus	Core 1 Focus	Core 2 Focus	Challenge Focus
	Build & Grow	**Cut, Cover, Compare**	**No Peeking!**	**Tenth Step**
Goal	Extend growing geometric patterns	Identify the recursive relationship within a growing pattern	Describe complex geometric patterns with words and numbers	Identify the functional relationship within a growing pattern
Materials	♦ Pattern card (see page 110) ♦ Multilink cubes	♦ Pattern cards (see page 110) ♦ Scissors ♦ 1″ Grid paper	♦ Pattern cards (see page 110) ♦ 1″ Color tiles ♦ Privacy folder	♦ Pattern cards (see page 110)
Suggested Activity Directions	Ask students to choose one of the pattern cards. Students recreate each pattern by building each step with multilink cubes. Prompt discussion about how the pattern grows with each step they build. Ask students to build the figure which would come next in the pattern. Provide time for students to try different pattern cards. 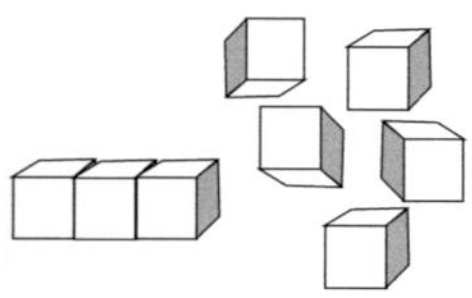	Provide students with one of the pattern cards. Ask them to cut out each of the figures within the pattern sequence. Students compare each step by overlapping the pieces in the correct order and describing the change from one step to the next. Lastly, students show the next step in the pattern using the grid paper. 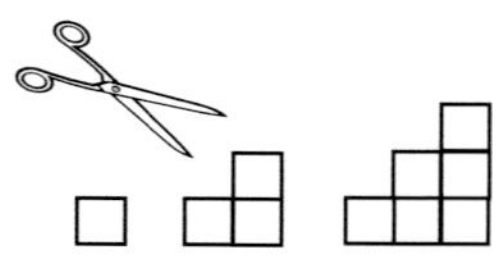	Separate partners by a privacy folder. The first student chooses a pattern card and describes the figures within the pattern. The partner follows the instructions and builds the figures using the color tiles. The partner may ask questions of the student describing the pattern. When finished, the student reveals the pattern card and partners compare the card to the finished product.	Distribute copies of the pattern cards. Explain to students that they are to determine the number of tiles needed to construct the tenth figure in the growing geometric pattern. Help students do this without building every single figure. Facilitate discussion when students explore the functional relationship between the step number and the number of tiles needed to continue the pattern. *2x step # +1*
Questions to Assess	♦ How many cubes do you need to build the first step? ♦ Are the number of cubes increasing or decreasing each time?	♦ How is the second step different from the first? ♦ What pattern do you notice as you build each step?	♦ How can you describe the position of the tiles? ♦ How many steps did you need to describe before your partner could predict the pattern?	♦ How does the step number relate to the number of tiles used? ♦ How many tiles would be in the tenth figure? How do you know your answer is correct?

CCSS Algebraic Thinking: Functions

Traditional Formative Assessment

Look at the function table.

Step	1	2	3	4
Visual	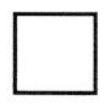	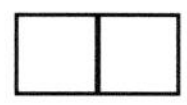	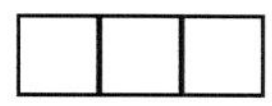	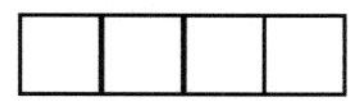
Toothpicks	4 toothpicks	7 toothpicks	10 toothpicks	13 toothpicks

How many toothpicks would be used in Step # 7?

- ○ 7 toothpicks
- ○ 22 toothpicks
- ○ 23 toothpicks
- ○ 28 toothpicks

Limitations: Any response, right or wrong, fails to reveal student thinking. A correct answer fails to distinguish whether a student recognizes the recursive relationship or the functional relationship. This is an important distinction when planning instruction.

Enhanced Formative Assessment

Look at the function table.

Step	1	2	3	4
Visual	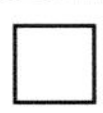	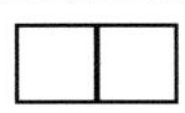	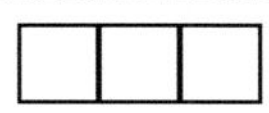	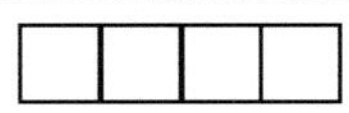
Toothpicks	4 toothpicks	7 toothpicks	10 toothpicks	13 toothpicks

How many toothpicks would be used in Step # 7?

Show your work.

Isaura

Look at the function table.

Step	1	2	3	4
Visual	□	□□	□□□	□□□□
Toothpicks	4 toothpicks	7 toothpicks	10 toothpicks	13 toothpicks

How many toothpicks would be used in Step # 7? >13

Show your work.

I don't now.

Isaura uses her knowledge of growing patterns to communicate that the value would be >13 since 13 toothpicks were used for step 4. However, she does not attempt to explore the relationship among the values.

Instructional Focus: **Rebuild**

Joanna

Look at the function table.

Step	1	2	3	4
Visual	□	□□	□□□	□□□□
Toothpicks	4 toothpicks	7 toothpicks	10 toothpicks	13 toothpicks

How many toothpicks would be used in Step # 7?

Show your work.

5 6
13 + 3 = 16
+3 7
19 + 3 = (21)

Joanna recognizes the recursive pattern of adding 3 toothpicks with each step. She accurately extends the pattern to step 7. Joanna does not see the relationship between the step number and the number of toothpicks.

Instructional Focus: **Core 1**

Karl

Look at the function table.

Step	1	2	3	4
Visual	□	□□	□□□	□□□□
Toothpicks	4 toothpicks	7 toothpicks	10 toothpicks	13 toothpicks

How many toothpicks would be used in Step # 7? 23

Show your work.

It is 23 because step3 + step 4 = step 7, so 10 toothpicks + 13 toothpicks = 23 toothpicks.

Karl reasons that step 3 plus step 4 equal step 7. If he applies the same strategy to step 1 + step 2 = step 3, he may recognize his error and try a different approach.

Instructional Focus: **Core 2**

Luis

Look at the function table.

Step	1	2	3	4
Visual	□	□□	□□□	□□□□
Toothpicks	4 toothpicks	7 toothpicks	10 toothpicks	13 toothpicks

How many toothpicks would be used in Step # 7?

Show your work.

Each step needs 3 toothpicks plus 1 more to close the box.

7x3=21 plus 1 more = 22 toothpicks.

Luis analyzes the functional relationship between the step number and the number of toothpicks. He accurately calculates the number of toothpicks for any step in this pattern using this strategy.

Instructional Focus: **Challenge**

CCSS Algebraic Thinking: Functions

	Rebuild Focus	Core 1 Focus	Core 2 Focus	Challenge Focus
	Toothpick Time	**Jumping Ahead**	**Why Not?**	**Another View**
Goal	Describe the recursive relationship in a function table	Explore the functional relationship between the step number and the number of elements	Describe and justify a functional relationship	Graph a functional relationship
Materials	♦ Enhanced formative assessment ♦ Toothpicks	♦ Enhanced formative assessment ♦ Toothpicks	♦ Enhanced formative assessment ♦ Toothpicks	♦ Enhanced formative assessment ♦ Toothpicks ♦ Graph paper
Suggested Activity Directions	Ask students to build each step of the pattern from the Enhanced Formative assessment using toothpicks. As students work, prompt them to identify the number of toothpicks used and how the step differs from the previous step. Provide time for students to continue the pattern through the tenth step and create new patterns. 	Discuss with students about the challenges of continuing a pattern to the 50th or 100th step. Explain that it is more efficient to find the relationship between the step number and the number of toothpicks needed. Explore the relationship between 1 and 4, 2 and 7, 3 and 10, and 4 and 13. Relate the rule to the picture. Ask students to build additional toothpick patterns and identify the rule. **3n + 1**	Pose the following solution for the Enhanced Formative Assessment problem: Since 1 + 3 = 4, could I add 4 toothpicks from step 1 to the 10 toothpicks in step 3 to get the correct number of toothpicks for step 4? Provide time for students to consider and explore this proposal and justify why it does not work. Prompt them to come up with a rule that does work. 	Ask students to graph the functional relationship from the Enhanced Formative Assessment. The *x*-axis is labeled with the step number and the *y*-axis shows the number of toothpicks used. Provide time for students to plot the points and discuss the rule as revealed on the graph. Students may create additional patterns with toothpicks to graph. 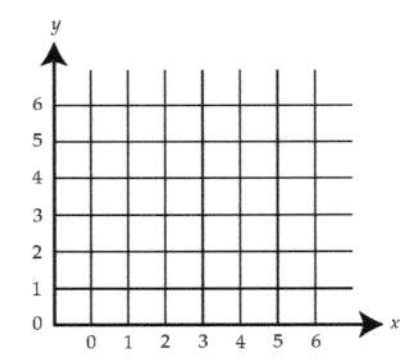
Questions to Assess	♦ How has the number of toothpicks changed from the last step? ♦ How many toothpicks do you predict will be needed for the next step?	♦ How can you find the number of toothpicks needed without building each step? ♦ What is the rule for the pattern you build?	♦ Does the rule work in all steps? ♦ How can you prove you have the correct number of toothpicks without building each step?	♦ What scale would be appropriate for the *y*-axis? ♦ What is the rule describing the relationship between the step number and the number of toothpicks?

Traditional Formative Assessment

Label each equation or inequality as True (T) or False (F)?

$20 - 6 = 10 + 10$	$17 - 4 < 13$	If $2+m=10$, then $m = 8$
$42 - 12 = 6 \times 5$	$29 > 2 \times 9$	If $28 = 7a$, then $a = 21$

Limitations: While the True/False format is more easily scored, it provides little insight into students' current knowledge of equations and inequalities. Students need to be asked to explain or justify their answers.

Enhanced Formative Assessment

Is each correct? Why or why not?

$42 - 12 = 6 \times 5$	$17 - 4 < 13$	If $28 = 7a$, then $a = 21$

Monique

Is each correct? Why or Why not?

42 - 12 = 6 x 5	17 - 4 < 13	If 28 = 7*a*, then *a* = 21
Can't rite it that way	Need =s	X

All wrong

Monique shares what she perceives to be incorrect with each problem. She does not yet understand that the equal sign is not always followed by the "answer."

Instructional Focus: **Rebuild**

Nate

Is each correct? Why or Why not?

no	no	don't know
42 - 12 = 6 x 5	17 - 4 < 13	If 28 = 7*a*, then *a* = 21
30 shood be the anser.	17 -4 13 is 23 no	?

Nate demonstrates his computation abilities. He recognizes the inequality as incorrect. Nate needs more experience with equations, which include operations on both sides of the equal sign.

Instructional Focus: **Core 1**

Omar

Is each correct? Why or Why not?

√	X	✓
42 - 12 = 6 x 5	17 - 4 < 13	If 28 = 7*a*, then *a* = 21
30 30 =	13 is not less than 13.	28 = 7 + 21

Omar proves the accuracy of the first equation and effectively explains why the inequality is not correct. However, he misinterprets the value of the variable "a" in 28 = 7*a*.

Instructional Focus: **Core 2**

Portia

Is each correct? Why or Why not?

42 - 12 = 6 x 5	17 - 4 < 13	If 28 = 7*a*, then *a* = 21
30 = 30	13 = 13	28 = 7 x 4 a = 4
yes	no	no

Portia understands how to read and interpret equations and inequalities. She accurately uses substitution to verify the inaccuracy of the variable in the equation 28 = 7*a*.

Instructional Focus: **Challenge**

CCSS Algebraic Thinking: Variables, Equations, & Inequalities

	Rebuild Focus **One Part at a Time**	Core 1 Focus **Make Them Equal**	Core 2 Focus **Alphabet Soup**	Challenge Focus **If This...Then That**
Goal	Identify and interpret the parts of equations and inequalities	Create and solve equations with operations on both sides of the = sign	Substitute values for variables	Explore equations and inequalities with 2 variables
Materials	♦ Numeral tiles (0–9) ♦ Operation tiles (+−×÷) ♦ Index card with = on side #1 and > on side #2	♦ Playing cards (minus face cards) ♦ Icosahedron die (1–20)	♦ Numeral tiles (1–6) ♦ Dry erase board and markers	♦ Numeral tiles (0–9) ♦ Dry erase board and markers
Suggested Activity Directions	Use the tiles and index card to make equations and inequalities. Help students focus on one side of the index card at a time to evaluate each expression (cover one side with a sheet of paper). Then direct students to focus on the complete equation or expression and indicate whether it is true or not true. If it is not true, ask them to flip or rotate the index card to make it true. 8 < 3 × 2	Use playing cards to create 2 equal quantities using any operation. Roll the die to determine the desired quantity. Choose from the cards to create 2 quantities equal to the value on the die. Write an equation to represent the cards. $6 \times 3 = 7 + 10 + 1$ $10 \div 1 = 5 \times 2$ 	Ask students to write the following equations and inequalities on a dry erase board: $28 = 7a$ $B + 4 = 6 + c$ $100 \div d = 50$ $e < f$ Challenge students to substitute a numeral tile for each variable, using each tile only once. Students may create their own puzzles, substituting variables for the values 1–6.	Provide an equation or inequality with two variables. Students explore how the value of one variable affects the value of the second. For example: $2a = 4 + b$ Students record the equation and use the tiles to substitute a value for "a" and then determine the correct value for "b." Discuss how changing the value of one variable affects the other. Students may create a table. a | b 2 | 0 3 | 2
Questions to Assess	♦ What is the quantity on each side of the index card? ♦ If the equation or inequality is not true, how could you change the index card to make it true?	♦ Will you use addition, subtraction, multiplication, or division? ♦ Is there another way to make the same quantity?	♦ Which variables were easiest to solve? Why? ♦ Is there more than one solution for each variable? ♦ Does the order in which you solve them make a difference?	♦ How does the value of one variable affect the value of the second variable? ♦ How might the relationship look on a graph?

Traditional Formative Assessment

Which sets of coins equal 50¢? Check (√) all that apply.

- O 1 quarter, dime, nickel
- O 2 quarters
- O 4 dimes, 4 nickels
- O 6 nickels, 2 dimes
- O 1 quarter, 2 dimes, nickel
- O 10 nickels
- O 1 quarter, 6 nickels
- O 3 dimes, 5 nickels

Limitations: Students may identify combinations totaling 50¢ but have not applied any algebraic or reasoning skills within a problem solving situation.

Enhanced Formative Assessment

The piggy bank held 5 coins with a value of 50¢.
What coins could be in the bank?

Quinton

The piggy bank held 5 coins with a value of 50¢. What coins could be in the bank?

The bank has pennys.

Quinton does not use any information from the problem to reach a solution when naming a type of coin the bank may hold.

Instructional Focus: **Rebuild**

Robert

The piggy bank held 5 coins with a value of 50¢. What coins could be in the bank?

(25) (25) 50¢

Robert has found a coin combination equaling 50¢ but does not recognize the need for the answer to include 5 coins.

Instructional Focus: **Core 1**

Sonja

The piggy bank held 5 coins with a value of 50¢. What coins could be in the bank?

(d) (d) (d) (d) (d)

10 20 30 40 50

Sonja correctly answers the problem by showing 5 coins with a value of 50¢.

Instructional Focus: **Core 2**

Tanya

The piggy bank held 5 coins with a value of 50¢. What coins could be in the bank?

Q	D	N	P
1	1	3	
	5		

Tanya finds 2 possible solutions to the problem. She uses a table to find possible solutions.

Instructional Focus: **Challenge**

CCSS Algebraic Thinking: Problem Solving & Reasoning

<table>
<tr><th></th><th>Rebuild Focus</th><th>Core 1 Focus</th><th>Core 2 Focus</th><th>Challenge Focus</th></tr>
<tr><th></th><th>This Little Piggy Has…</th><th>Framing the Problem</th><th>Count the Ways</th><th>How Many Ways?</th></tr>
<tr><th>Goal</th><td>Understand the problem</td><td>Represent the problem</td><td>Search for more than one solution</td><td>Solve more complex problems</td></tr>
<tr><th>Materials</th><td>♦ Coins
♦ 4″×6″ index cards
♦ Markers</td><td>♦ Blank paper
♦ Coins
♦ Markers</td><td>♦ Index cards
♦ Coins</td><td>♦ ½″ Graph paper</td></tr>
<tr><th>Suggested Activity Directions</th><td>Ask each student to draw a piggy bank on the index card and secretly place 3 coins under the card. Students ask questions about the coins to determine the hidden combination under a student's card. Questions may be about the attributes of the coins (e.g., value, color, size), but not the names of the coins. Questions continue until coins and value are correctly identified.
</td><td>Use the piggy bank problem. Ask each student to draw a large piggy bank. Reread the problem to reinforce that the bank contains exactly 5 coins. Ask students to draw 5 large circles within the piggy bank. Students may place a single coin within each circle and count the total to see if the combination equals 50¢. Help students record equations equaling 50¢.
</td><td>Prompt students to consider if there might be another answer to the piggy bank problem. Suggest that they work together to find all the possible combinations for 50¢ using quarters, dimes, and nickels. Each solution should be placed on an index card using the actual coins. Next, ask students to determine which solutions included five coins and how the cards could be organized.
50 cents</td><td>Ask students to find all the possible coin combinations equaling 50¢. To extend students' thinking further, ask them to consider how the inclusion of pennies changes the number of possibilities. Encourage students to use an organized table to show the combinations found.
<table>
<tr><th>Q</th><th>D</th><th>N</th><th>P</th></tr>
<tr><td>2</td><td>0</td><td>0</td><td>0</td></tr>
<tr><td>1</td><td>2</td><td>1</td><td>0</td></tr>
<tr><td>1</td><td>1</td><td>3</td><td>0</td></tr>
<tr><td>1</td><td>1</td><td>2</td><td>5</td></tr>
<tr><td>1</td><td>1</td><td>1</td><td>10</td></tr>
<tr><td>1</td><td>1</td><td>0</td><td>15</td></tr>
<tr><td>1</td><td>0</td><td>5</td><td>0</td></tr>
<tr><td>1</td><td>0</td><td>4</td><td>5</td></tr>
</table></td></tr>
<tr><th>Questions to Assess</th><td>♦ Is one coin worth more than 20¢?
♦ Are any of the coins the same?
♦ Is the total worth less than 40¢?</td><td>♦ How does the picture represent the problem?
♦ How can we keep track of the combinations we already tried?</td><td>♦ How do we know if we have found all the possible ways to make 50¢?
♦ How could we organize the index cards?</td><td>♦ What patterns do you notice in the organized list?
♦ Are the patterns helpful? Why or why not?</td></tr>
</table>

Algebraic Thinking: Rates of Change

Traditional Formative Assessment

Match each graph with the correct situation description.

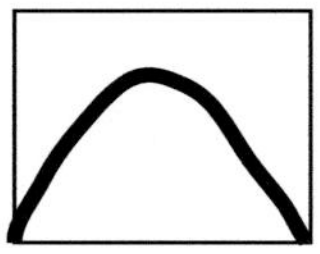

○ Growth of a plant over two weeks

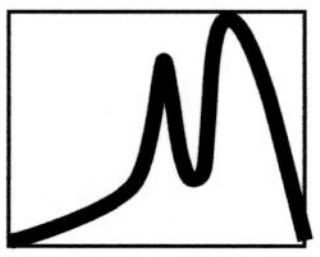

○ Filling and emptying a glass with water

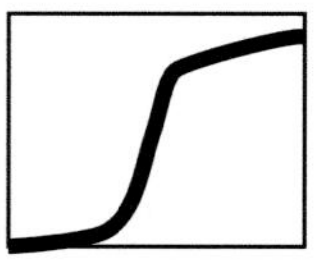

○ The speed of a roller coaster

Limitations: There is a possibility of more than one answer for any given prompt. Almost any answer could be justified through a description.

Enhanced Formative Assessment

Use the graph to describe sales at the lemonade stand through the day.

Sales

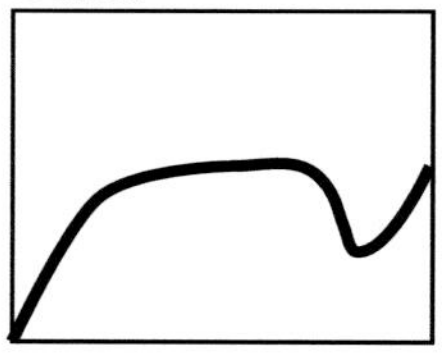

Sketch the graph to show how temperature changes from the middle of the day to the evening. Explain why you drew the graph the way you did.

Temperatures

Algebraic Thinking: Rates of Change

Ulma

Use the graph to describe sales at the lemonade stand through the day.

They sold a lot of lemonade.

Sales

Sketch the graph to show how temperature changes from the middle of the day to the evening. Explain why you drew the graph the way you did.

It is hot.

Temperatures

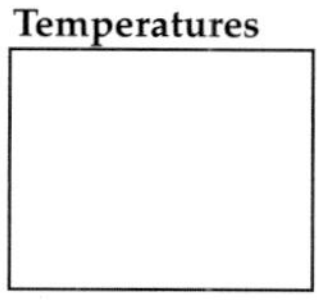

Ulma restates what she understands from the problem. However, she is unable to relate each situation to a graphical representation.

Instructional Focus: **Rebuild**

Vernon

Use the graph to describe sales at the lemonade stand through the day.

The line goes up and down and up again. I think next it will go back down.

Sales

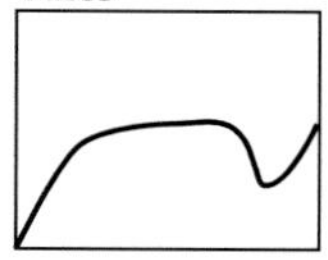

Sketch the graph to show how temperature changes from the middle of the day to the evening. Explain why you drew the graph the way you did.

The line could go up or down.

Temperatures

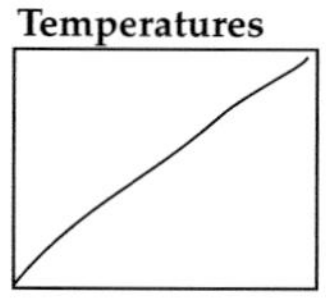

Vernon describes the changes in the line but fails to connect the changes in the line to the variables within the scenario.

Instructional Focus: **Core 1**

Wayne

Use the graph to describe sales at the lemonade stand through the day.

They sold more and more lemonade. Then they sold less and then more til they stopped.

Sales

Sketch the graph to show how temperature changes from the middle of the day to the evening. Explain why you drew the graph the way you did.

The temperature goes up and then back down in the evening.

Temperatures

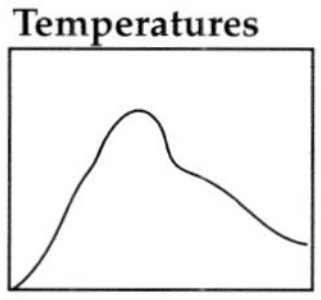

Wayne correctly describes the rate change but assumes that graphs always start with an increase.

Instructional Focus: **Core 2**

Yohanna

Use the graph to describe sales at the lemonade stand through the day.

Lemonade sales went up then stayed the same for a while then went down and then back up again.

Sales

Sketch the graph to show how temperature changes from the middle of the day to the evening. Explain why you drew the graph the way you did.

First it is hot and then it gets cooler to night time.

Temperatures

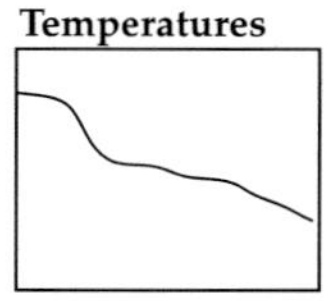

Yohanna is able to accurately describe and sketch the rates of change.

Instructional Focus: **Challenge**

Algebraic Thinking: Rates of Change

	Rebuild Focus	Core 1 Focus	Core 2 Focus	Challenge Focus
	Talk It Out	**What is Changing?**	**Ups and Downs**	**Constant or Variable**
Goal	Connect rates of change to a graphic representation	Identify the attributes when viewing rates of change	Analyze graphs with increasing and decreasing rates of change	Collect data and display the rate of change
Materials	♦ Dry erase boards and markers	♦ Dry erase boards and markers ♦ Graph paper	♦ Graph paper ♦ Large index cards	♦ Graph paper
Suggested Activity Directions	Present students with familiar scenarios demonstrating rates of change. Ask students to describe the changes (e.g., number of students entering school in the morning, spending money at the mall, speed of a roller coaster). Provide time for students to sketch the changes as they consider the incremental changes. Focus on vocabulary such as increasing, decreasing, constant, and time. 	Ask students to name things that change over time. Provide examples such as: height, the amount of rainfall in a month, the growth of a flower, and money in a bank. Ask students to discuss how these scenarios change over time. Demonstrate how students can sketch such changes as the attributes increase, decrease, or stay the same. When students are ready, they may transfer their sketches to graph paper. 	Share scenarios in which quantities might increase or decrease (e.g., temperature—morning to noon or noon to evening; allowance-spending; eating cookies from a cookie jar). Ask students to sketch a graph for each scenario. Students then describe each graph on an index card. Shuffle graphs and cards and work as a group to match correct descriptions with correct graphs. 	Ask students to generate a list of data scenarios they could collect which would demonstrate a change over time (e.g., students ordering school lunch over a week's time, class attendance Monday to Friday, money spent at the school store, height of plants). Students simulate the scenario, collect, and display the data on graph paper. Help students define the change as constant or variable. 
Questions to Assess	♦ What if there are more students on the next bus? ♦ How is the amount of money changing as I buy more items? ♦ Is the roller coaster moving at a constant speed?	♦ If the amount of money in the bank decreases, how is that decrease represented with the line? ♦ What will the graph look like if your height stays the same for a long time?	♦ Which vocabulary helped you match the description to the correct graph? ♦ How can you make your description specific enough so it only matches a single graph?	♦ Could you predict the rate of change for another time interval? ♦ What attribute is changing over time? ♦ How would you describe the degree of the change?

4

Geometry

coordinates

congruency

two-dimensional

line

symmetry

transformations

rotational

three-dimensional

Traditional Formative Assessment

Name each quadrilateral.

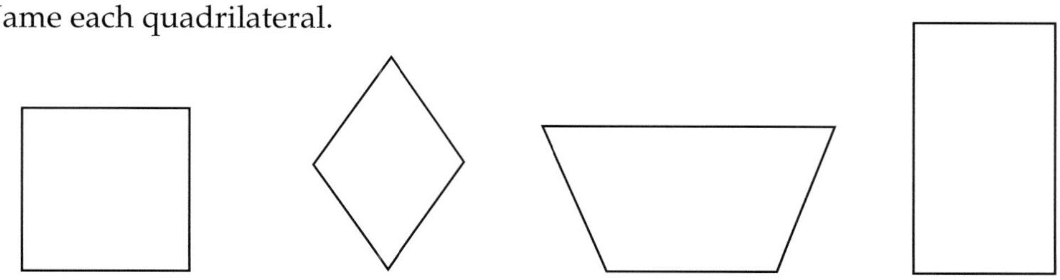

Limitations: Naming each quadrilateral does not assess whether students recognize the attributes of each, nor does it address the classification of quadrilaterals.

Enhanced Formative Assessment

Describe the attributes of each quadrilateral and draw an example.

- square
- rectangle
- rhombus
- trapezoid

Aja

Describe the attributes of each quadrilateral and draw an example.

- Square 4 sides all the same
- Rectangle 2 short sides and 2 long sides
- Rhombus
- Trapezoid ?

Aja understands that squares and rectangles have 4 sides, but does not mention the square corners or parallel lines. She assumes all rectangles must have short and long sides.

Instructional Focus: **Rebuild**

Brent

Describe the attributes of each quadrilateral and draw an example.

- Square 4 equal sides + 4 square Corners
- Rectangle 4 sides + 4 square Corners
- Rhombus a square but turned
- Trapezoid a square with a long bottom

Brent describes the square and rectangle correctly. However, he erroneously thinks that position defines a rhombus and a trapezoid is a square.

Instructional Focus: **Core 1**

Cory

Describe the attributes of each quadrilateral and draw an example.

- Square 4 sides same length, 4 square corners
- Rectangle same as square ~~but~~ sides can be different
- Rhombus 4 sides same lengh, no square corners
- Trapezoid the red pattern block

Cory accurately describes the square and rectangle, but inaccurately states that a rhombus cannot have square corners. He also fails to describe the sides and angles of the trapezoid.

Instructional Focus: **Core 2**

Dreanna

Describe the attributes of each quadrilateral and draw an example.

- Square 4 = sides, 4 right ∠s
- Rectangle 2 sets paralel lines, 4 right ∠s
- Rhombus 4 = sides
- Trapezoid 4 sides but one set is paralel

Dreana describes all of the quadrilaterals correctly. Additionally, her examples demonstrate her understanding that squares are subset of rectangles.

Instructional Focus: **Challenge**

CCSS **Geometry: Two-Dimensional Shapes**

<table>
<tr><th></th><th>Rebuild Focus</th><th>Core 1 Focus</th><th>Core 2 Focus</th><th>Challenge Focus</th></tr>
<tr><th></th><th>Shape Puzzle</th><th>Which Has What?</th><th>That's Mine!</th><th>It's All in a Name</th></tr>
<tr><th>Goal</th><td>Name and describe quadrilaterals</td><td>Compare quadrilaterals</td><td>Sort shapes according to attributes</td><td>Construct and classify quadrilaterals</td></tr>
<tr><th>Materials</th><td>◆ Cutouts of rectangles (including squares), rhombuses and trapezoids
◆ Chart paper and markers
◆ Scissors</td><td>◆ Geoboards and geobands
◆ Dry erase boards and markers</td><td>◆ Geoboards and geobands</td><td>◆ Drawing paper
◆ Rulers
◆ Protractors</td></tr>
<tr><th>Suggested Activity Directions</th><td>Distribute the shape cutouts to students. Ask students to take turns naming and describing the shapes (by sides and angles). Record responses on a table. When the table is complete, cut along the lines to create a puzzle. Request that students work together to reassemble the table, aligning the correct attributes for each shape.
<table><tr><th>Name</th><th>Sides</th><th>Angles</th></tr><tr><td>Square
</td><td>4, all same length, parallel (2 sets)</td><td>4, square or right</td></tr><tr><td></td><td></td><td></td></tr><tr><td></td><td></td><td></td></tr><tr><td></td><td></td><td></td></tr></table></td><td>Partners make a trapezoid on one geoboard and a square on the other. Prompt students to draw intersecting loops on the dry erase board and record how the shapes are alike and how they are different. Encourage students to consider the attributes of the sides and angles in their comparisons. Rotate the geoboard and ask students if the attributes have changed. Repeat with a nonsquare rhombus and a square.
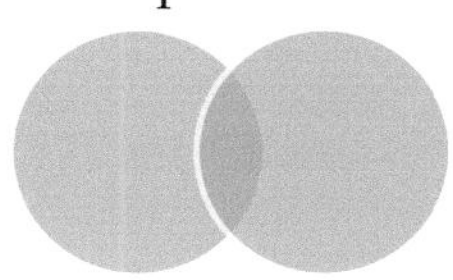</td><td>Ask students to make a quadrilateral on their geoboards. Allow each to share the name of their shape. Take time to compare shapes with the same name that look different. Next, position the geoboards flat on each students' lap. Call out a specific attribute. If a student's shape fits that attribute, he should hold up his geoboard to show the group and call out "That's Mine!" Such attributes may include one set of parallel lines, all right angles, all sides equal length, no right angles, etc.
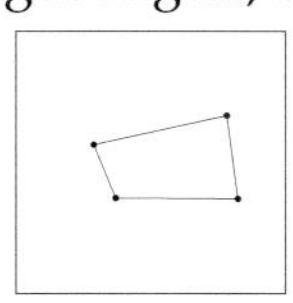</td><td>Distribute tools to students and allow them to draw several quadrilaterals. Explain that they should draw 2 different trapezoids, 2 different rhombuses, 2 different squares, and 2 different rectangles. Then ask students to label each shape with all of the appropriate names. For example, a square could be labeled as a square, rectangle, rhombus, parallelogram, and quadrilateral.
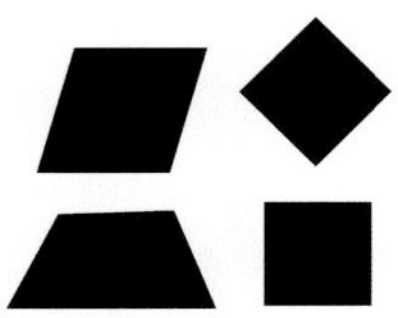</td></tr>
<tr><th>Questions to Assess</th><td>◆ What can you say about the sides of a trapezoid?
◆ Does the shape have any other names?</td><td>◆ What do the shapes have in common?
◆ How are the angles different?</td><td>◆ Could two rhombuses look different? How?
◆ Could this shape be called by another name?</td><td>◆ Does that rectangle have another name?
◆ What specifically makes that quadrilateral a rhombus?</td></tr>
</table>

Traditional Formative Assessment

Match the name of each shape with the correct shape.

- Cylinder
- Triangular Pyramid
- Cone
- Rectangular Prism

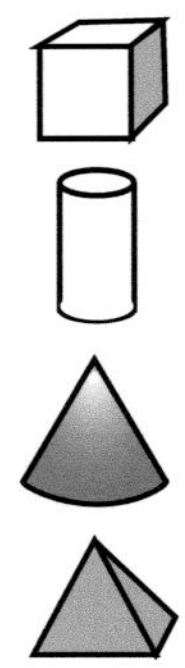

Limitations: Asking a student to simply name shapes fails to measure a student's ability to describe and classify three-dimensional figures.

Enhanced Formative Assessment

Name and describe each 3-dimensional shape.

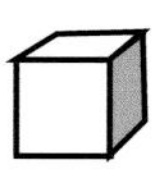

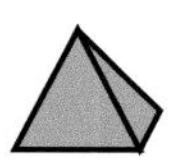

CCSS Geometry: Three-Dimensional Shapes

Elijah

Name and describe each 3-dimensional shape.

box -

?

cone

Elijah correctly names the cone and is unable to describe any of the three-dimensional shapes.
Instructional Focus: **Rebuild**

Fran

Name and describe each 3-dimensional shape.

cube - like a box

pyramid - a point on top

cone - a point on top

Fran names the shapes and attempts to describe each. Her descriptions are simplistic and lack geometric vocabulary.
Instructional Focus: **Core 1**

Girard

Name and describe each 3-dimensional shape.

cube - 6 squares together

pyramid - triangles for faces

cone - a circle base and a point

Girard names each shape and makes reference to the shapes of the faces using some math vocabulary.
Instructional Focus: **Core 2**

Haley

Name and describe each 3-dimensional shape.

rectangular prism, cube - 6 rectangel faces and 12 edges

square pyramid - a square base with 4 triangle faces and 8 edges

a circle for the base and a point at the top

Haley accurately identifies all of the three-dimensional shapes. Her descriptions include rich math vocabulary.
Instructional Focus: **Challenge**

CCSS Geometry: Three-Dimensional Shapes

	Rebuild Focus	Core 1 Focus	Core 2 Focus	Challenge Focus
	Which Is It?	**Property Loops**	**Draw, Cut, & Fold**	**Build It!**
Goal	Identify and describe pyramids and prisms	Identify and sort pyramids and prisms	Make nets of prisms and pyramids	Construct more complex three-dimensional shapes
Materials	♦ Three-dimensional shapes including pyramids and prisms	♦ Various pyramids and prisms ♦ String ♦ Sticky notes	♦ Various pyramids and prisms ♦ Drawing paper ♦ Scissors ♦ Tape	♦ Straws ♦ Paperclips ♦ Scissors
Suggested Activity Directions	Display the shapes. Choose one of the shapes to describe. Ask students to identify which shape you described. When the correct shape is identified, discuss which clues were most helpful in the identification process. Also discuss how some shapes were eliminated by the clues. Invite students to take turns offering the descriptions so others may identify the correct shape. 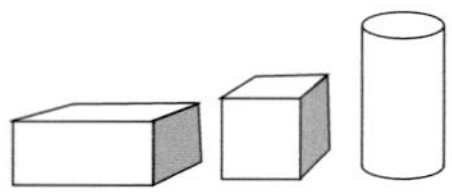	Use loops of string to make property loops. Label each loop with a sticky note. Invite students to place the three-dimensional shapes within the correct set(s). Ask students to justify the placement of each shape. Sample labels might include: ♦ rectangular base ♦ triangular faces ♦ congruent faces ♦ prisms 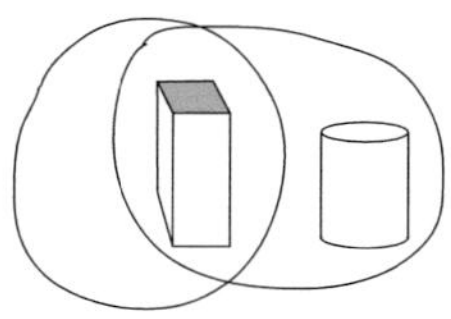	Model for students how to trace each face of the shape on the drawing paper. Demonstrate how to move the shape so that all faces can be traced with adjoining edges. Cut, fold, and tape the net and compare the original shape with the newly constructed shape. Invite students to choose a shape and make a net. Before constructing the shapes, ask the group to match each net to the appropriate shape. 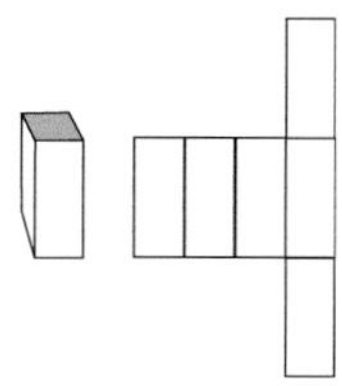	Demonstrate for students how to use the paperclips to connect the straws. Open the paperclip to an S shape and insert a straw on each end. The opening on each end of the paperclip may need to be adjusted to fit snuggly in the straw. The paperclip may then be bent as needed to form a vertex. Allow students to choose which shapes to construct (e.g., triangular prism using right triangles, trapezoidal prism, hexagonal pyramid). 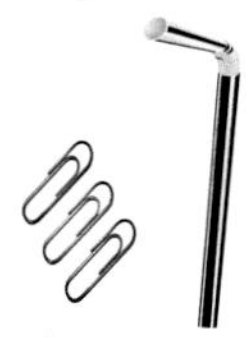
Questions to Assess	♦ What do we know about the faces of the shape? ♦ Is there another possible answer based on the clues?	♦ How many rectangular prisms are in the set of shapes? ♦ How are prisms and pyramids alike? How are they different?	♦ What strategy can you use to keep track of the faces you already traced? ♦ How do you know which shape the net will make?	♦ What are the shapes of the faces? ♦ How many straws (edges) does your shape include?

CCSS Geometry: Congruency

Traditional Formative Assessment

Circle the pair of congruent shapes.

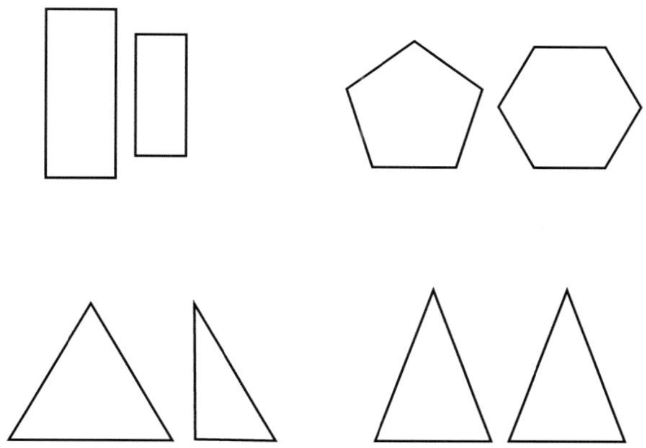

Limitations: Without an explanation of the answer choice, it is impossible to know if the students can identify and define congruent figures.

Enhanced Formative Assessment

Circle all sets of congruent shapes. Explain how you decided which shapes are congruent.

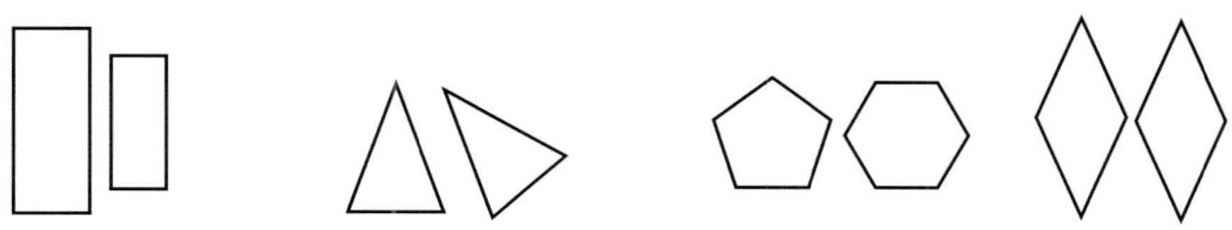

How do you know the shapes are congruent?

CCSS Geometry: Congruency

Illcia

Circle all sets of congruent shapes. Explain how you decided which shapes are congruent.

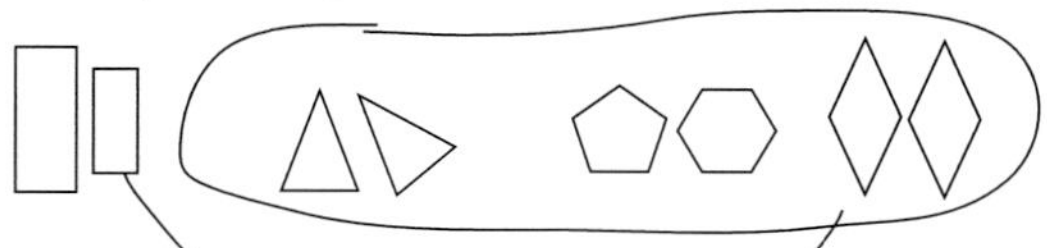

How do you know the shapes are congruent?

too small

same size

Illcia focuses only on the size of the shapes and erroneously identifies shapes of similar size as congruent.

Instructional Focus: **Rebuild**

Jorge

Circle all sets of congruent shapes. Explain how you decided which shapes are congruent.

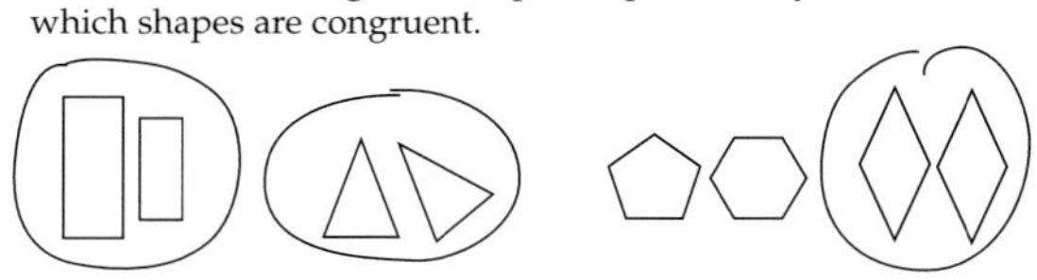

How do you know the shapes are congruent?

The shapes match.

Jorge inaccurately picks pairs of like shapes as congruent, regardless of size.

Instructional Focus: **Core 1**

Kenia

Circle all sets of congruent shapes. Explain how you decided which shapes are congruent.

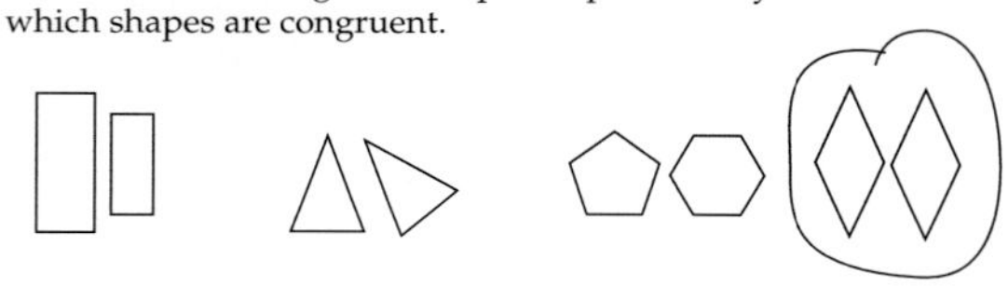

How do you know the shapes are congruent?

They are the same size and the same shape.

Kenia correctly identifies congruent figures as having the same size and same shape. However, she does not recognize differently positioned triangles as congruent.

Instructional Focus: **Core 2**

Leonard

Circle all sets of congruent shapes. Explain how you decided which shapes are congruent.

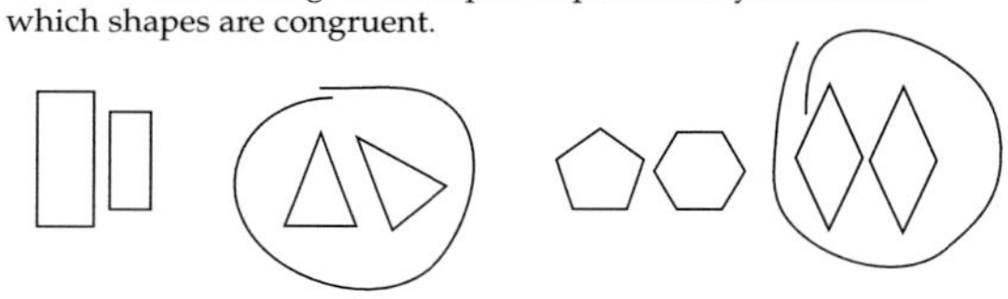

How do you know the shapes are congruent?

The shapes are the same size and shape even if they are turned.

Leonard correctly identifies the two sets of congruent shapes.

Instructional Focus: **Challenge**

CCSS Geometry: Congruency

	Rebuild Focus	Core 1 Focus	Core 2 Focus	Challenge Focus
	Cut for Congruency	**Large and Small**	**Pentominoes**	**Congruent Angles**
Goal	Change shape to make shapes congruent	Change size to make shapes congruent	Change in orientation maintains congruence	Congruent angles
Materials	♦ Attribute blocks ♦ Drawing paper ♦ Scissors	♦ Attribute blocks ♦ Drawing paper ♦ Scissors ♦ Rulers	♦ Square tiles ♦ 1″ Grid paper ♦ Scissors	♦ Attribute blocks ♦ Protractor ♦ Drawing paper
Suggested Activity Directions	Display 2 different large shapes. Invite students to share how the shapes are the same and how they are different. Trace and cut out the 2 shapes. Ask students how they could cut the shapes to make them congruent. Give students time to trace and make additional congruent shapes out of different shapes of the same size (e.g., small rectangle and small circle). 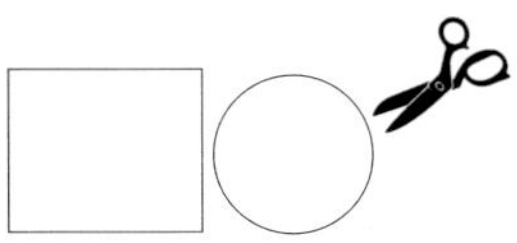	Display a large rectangle and a small rectangle of the same color and thickness. Define congruence and ask students if the shapes are congruent. Encourage students to explain why the shapes are not congruent. Allow students time to make congruent shape partners by sorting and tracing pairs of congruent shapes. Students may also draw original congruent shapes using the ruler. 	Ask students to make all of the different noncongruent arrangements for 5 square tiles. The only rule is that adjacent tiles must touch by a full edge rather than a single point. 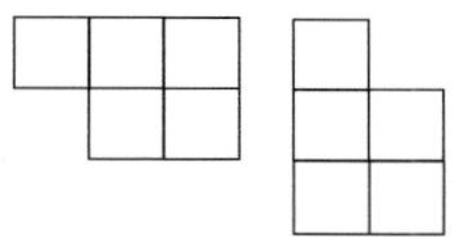 Students may cut out the shapes and overlap the pieces to check for congruence.	Students begin to explore the idea that noncongruent shapes can have congruent angles. Ask students to trace 2 attribute blocks of the same shape but different sizes. Model for students how to compare the angles among the 2 shapes. Students may use the drawing tools to make their own noncongruent shapes with congruent angles (similar shapes). 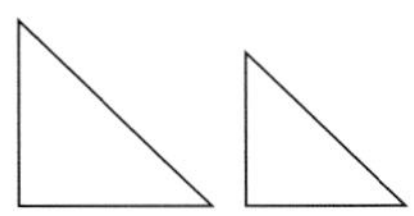
Questions to Assess	♦ How would you describe congruent shapes? ♦ What strategy could you use to check for congruence?	♦ How could you prove that the shapes are the same size? ♦ Can 2 shapes of the same color be congruent?	♦ How can flips, slides, and turns help you check for congruency? ♦ Did we find all 12 possible arrangements?	♦ Are all of the angles congruent? ♦ How do the lengths of the sides compare on these shapes?

Traditional Formative Assessment

What are the coordinates of the heart?

- O 6, 6
- O 6, 4
- O 4, 6
- O 4, 4

What are the coordinates of the smiley face?

- O 3, 3
- O 3, 0
- O 0, 6
- O 0, 3

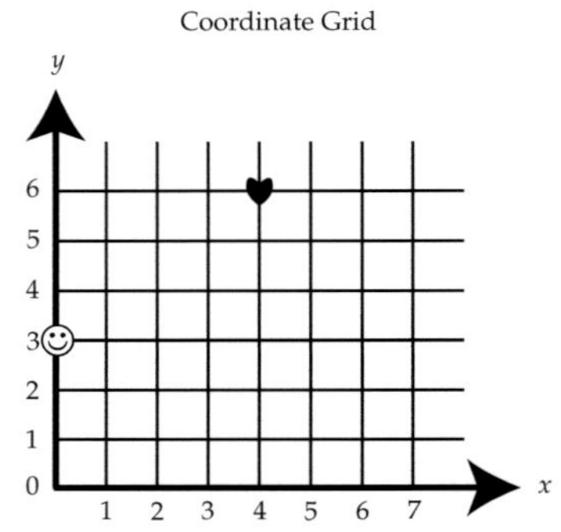

Limitations: A correct answer may be the result of understanding, or simply be a guess.

Enhanced Formative Assessment

Describe the location of the heart.

Draw a triangle at the coordinates **6, 4**.

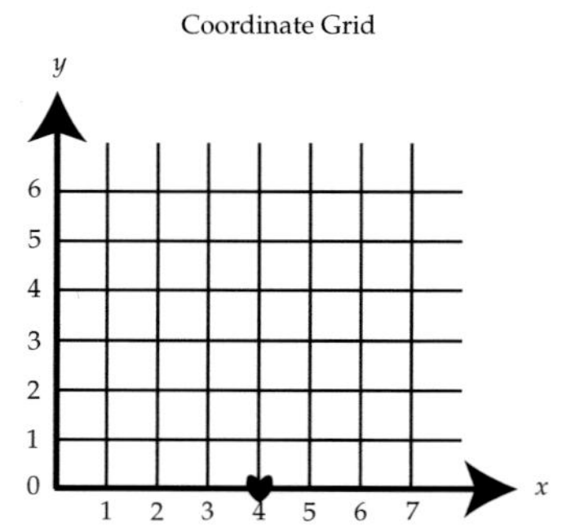

Micah Describe the location of the heart. *at the bottom* Draw a triangle at the coordinates **6**, **4**.	Naomi Describe the location of the heart. *on the 3* Draw a triangle at the coordinates **6**, **4**.
Micah describes the location of the heart as "at the bottom." He does not use the coordinates to identify a point. Instructional Focus: **Rebuild**	Naomi identifies the 3 as a coordinate for the heart. She places a triangle on every 6 and every 3 on the coordinate grid. Instructional Focus: **Core 1**
Ohare Describe the location of the heart. *0, 3* Draw a triangle at the coordinates **6**, **4**.	Pat Describe the location of the heart. *3, 0* Draw a triangle at the coordinates **6**, **4**.
Ohare correctly places the triangle at the designated coordinates, but inaccurately identifies the coordinates for the heart. Instructional Focus: **Core 2**	Pat understands how to accurately name and locate points on a coordinate grid. Instructional Focus: **Challenge**

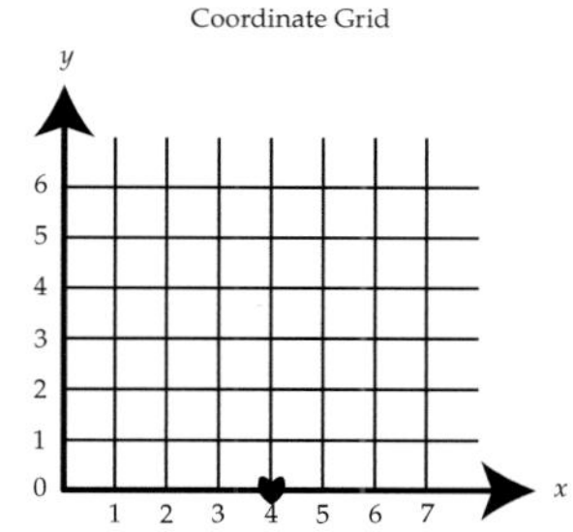

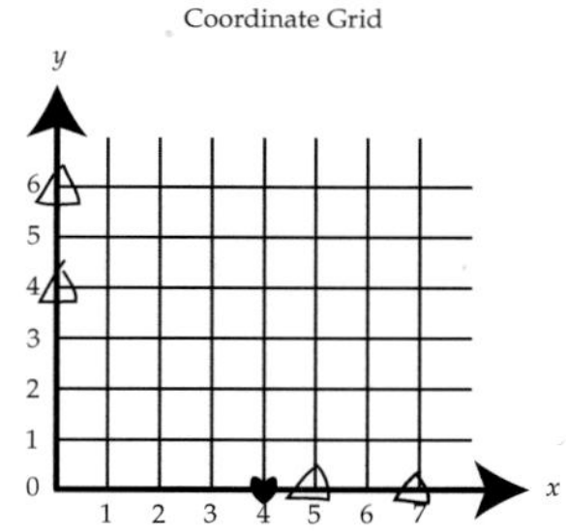

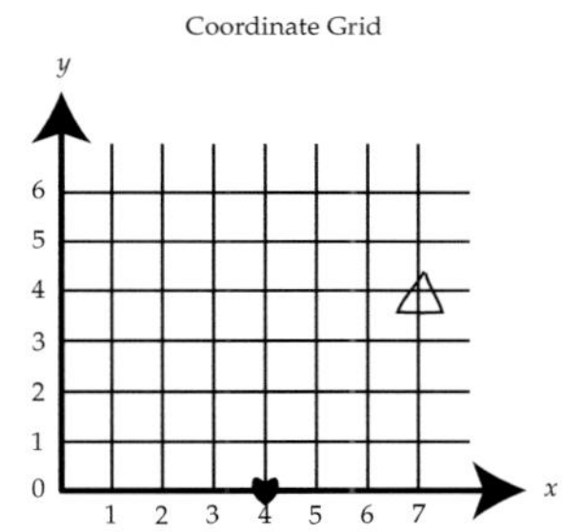

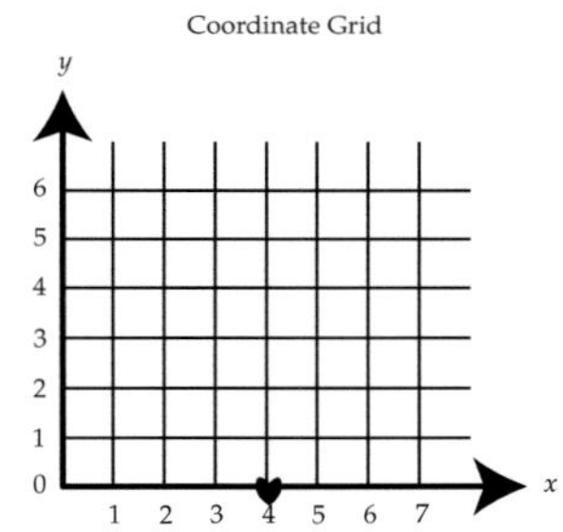

CCSS Geometry: Coordinate Systems

	Rebuild Focus	Core 1 Focus	Core 2 Focus	Challenge Focus
	Three in a Row	**Criss-Cross**	**Quick Draw**	**Extending the *x*-Axis**
Goal	Understand points are identified by coordinates	Locate the intersection of the *x* and *y* coordinates	Understand that the *x* coordinate is named first in a ordered pair	Explore negative coordinates
Materials	♦ Geoboards ♦ 1″ Paper squares ♦ Sticky dots	♦ Grid paper ♦ Pipe cleaners	♦ Large grid paper ♦ Playing cards (minus face cards) ♦ Pawns (2 colors)	♦ Grid paper
Suggested Activity Directions	Work with students to set up the geoboard as a coordinate grid. Place sticky dots along the bottom and left side to label each axes. Explain to students that we name the *x*-axis (over), followed by the *y*-axis (up). Play in teams, racing to get 3 in a row. A team names a point and marks the spot by poking a paper square over the nail. Play continues until a team gets 3 in a row for the win. 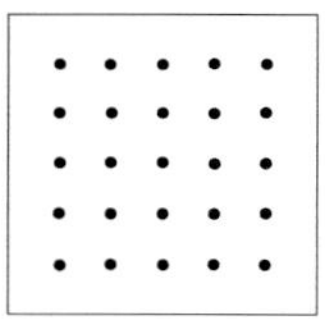	Draw the *x* and *y* axes on the grid paper for each student. Discuss the labels along each axis as students write the numerals on each line. Call out coordinates one at a time beginning with the x-axis. Students place the pipe cleaner on the appropriate value for each axis and mark the point of intersection with a dot. Provide time for students to then label each point with the correct coordinates. 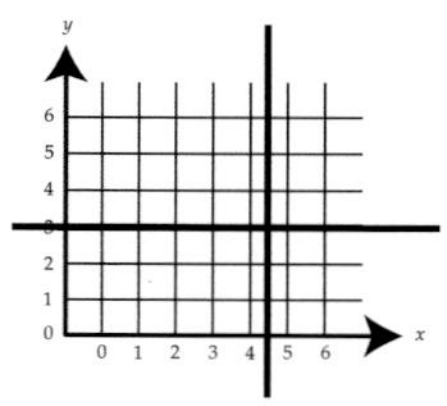	Draw and label the *x* and *y* axes on the large grid paper. Flip the first card to name the x coordinate and the second card to name the y coordinate. Ask students to identify the correct location for the pawn. Provide time for students to take turns flipping cards, plotting points, and labeling each point with the correct coordinates. Students may play on two teams to flip cards and race to be the first to find the correct point. The opposing team must agree the point is correctly identified before the next turn.	Starting at the center of the grid paper, demonstrate how to draw and label the upper right quadrant of the Cartesian coordinate graph with zero at the bottom left. Ask students to extend the *x* axis to the left. Discuss the value of the points to the left and label the negative integers. Provide time for students to find and record various coordinates by plotting symmetrical positions. 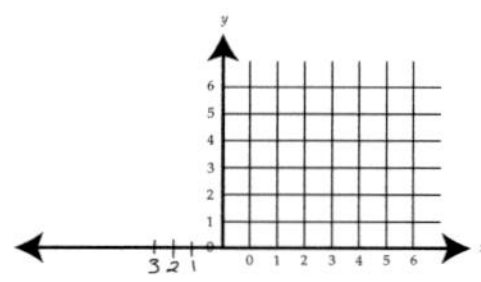
Questions to Assess	♦ What would the next coordinate be? ♦ How can I communicate the point our team wants to mark?	♦ What are the coordinates? ♦ Where do the pipe cleaners intersect?	♦ Which coordinate is named first? ♦ How far is the point from the origin?	♦ Where is the line separating the negative coordinates from the positive coordinates? ♦ Where is the origin?

Geometry: Transformations

Traditional Formative Assessment

Circle the word that correctly describes the transformation of the shape.

 translation rotation 90° reflection

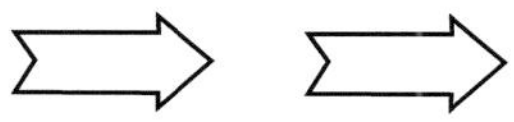 translation rotation 90° reflection

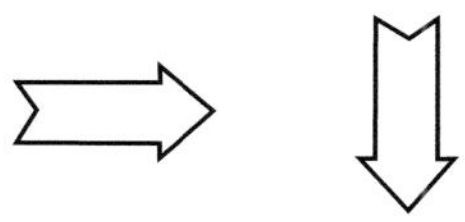 translation rotation 90° reflection

Limitations: Student-drawn representations help confirm understanding or offer evidence of a misconception or error pattern more comprehensively than do multiple choice responses.

Enhanced Formative Assessment

Draw the shape to show each transformation.

Show a translation of the shape. 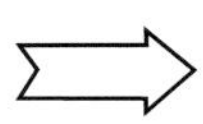

Show a reflection of the shape. 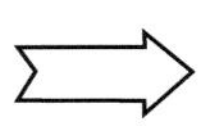

Show a 90° rotation of the shape. 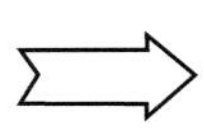

Geometry: Transformations

Quinn

Draw the shape to show each transformation.

Show a translation of the shape.

Show a reflection of the shape.

Show a 90° rotation of the shape.

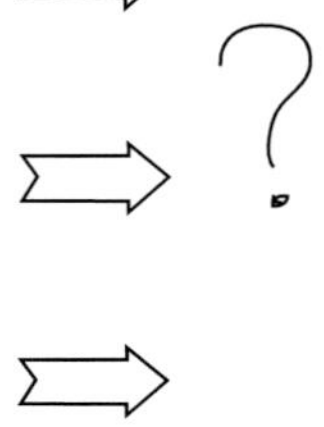

Quinn is unable to represent any of the transformations.

Instructional Focus: **Rebuild**

Ray

Draw the shape to show each transformation.

Show a translation of the shape.

Show a reflection of the shape.

Show a 90° rotation of the shape.

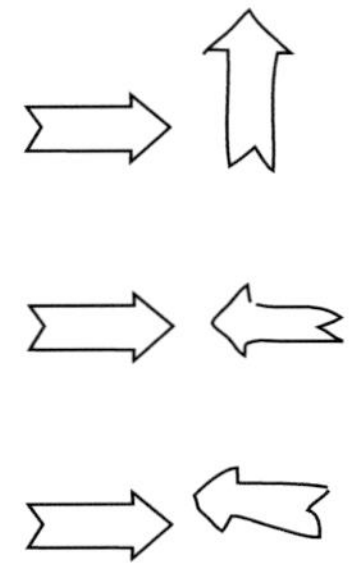

Ray accurately draws the reflection.

Instructional Focus: **Core 1**

Samone

Draw the shape to show each transformation.

Show a translation of the shape.

Show a reflection of the shape.

Show a 90° rotation of the shape.

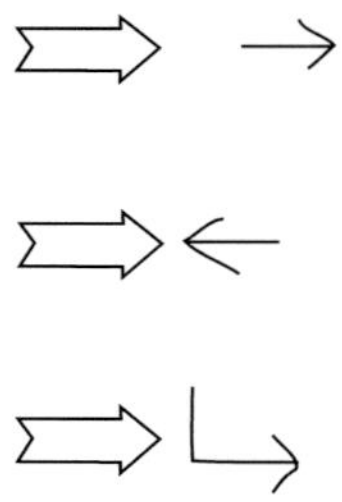

Samone draws the translation and refection correctly. She represents a rotation as a 90° angle.

Instructional Focus: **Core 2**

Tomas

Draw the shape to show each transformation.

Show a translation of the shape.

Show a reflection of the shape.

Show a 90° rotation of the shape.

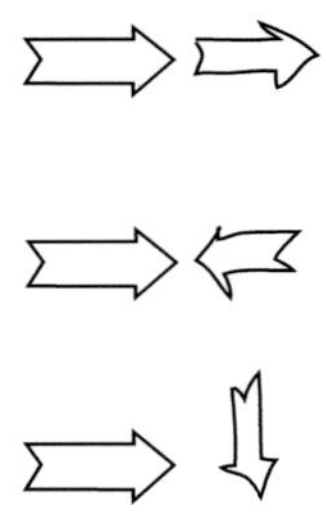

Tomas correctly draws each transformation.

Instructional Focus: **Challenge**

Geometry: Transformations

	Rebuild Focus	Core 1 Focus	Core 2 Focus	Challenge Focus
	Sliding Along	**Flipped!**	**This Way and That**	**Here to There**
Goal	Understand translations	Understand reflections	Understand rotations	Understand compositions (combination of 2 or more transformations)
Materials	♦ Large grid paper ♦ Small toy car (or other marker) ♦ Numeral cube (1–6)	♦ Grid paper (inch) ♦ Clear inch squares (cut from transparency) ♦ Marker	♦ Grid paper (inch) ♦ Multilink cubes	♦ Dot paper (see page 111)
Suggested Activity Directions	Label the upper left square of the grid paper with START and the lower right square with FINISH. Place the car on START and explain to students the need to get to the finish line using translations (slides). Students roll the die to indicate the number of translations. Discuss the directions of the translations (e.g., left, right, up, down, north, south, east, west). Students take turns translating the car until the car reaches the FINISH line. 	Students use markers to draw a design on a clear square. Model translations and reflections for students on grid paper using the clear inch design. Students draw each design in the new location and label the transformation as a translation or reflection. Remind students to describe the direction of each transformation (e.g., reflection to the right). 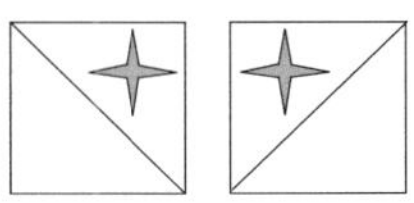	Allow students to build a shape using 6 multilink cubes. Model a rotation and explain the amount of the rotation must be defined (e.g., 90°, 180°, 360°). Instruct students to place the figure on the grid paper and demonstrate translations, reflections, and rotations. Students record and label each transformation. Remind students to include the direction or degree of rotation for each move. 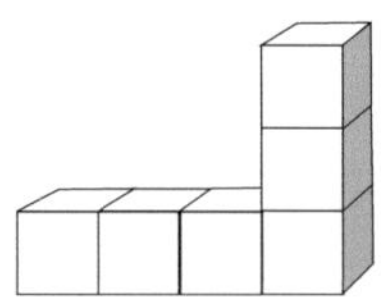	Draw a U shape in one corner of the dot paper. Ask students to draw a congruent U shape at another location on the dot paper away from the original shape. Students then describe the combination of transformations needed to get from the first U shape to the second U shape. Students label each transformation on the dot paper. 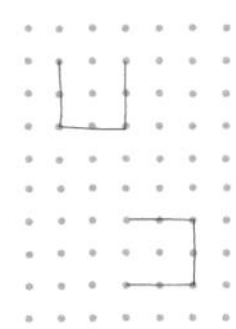
Questions to Assess	♦ How many translations did we take on this turn? ♦ In which direction did the car translate?	♦ How do you know you made a reflection? ♦ How is a reflection different from a translation?	♦ Is it possible for a reflection and a rotation to create the same result? ♦ How would you describe each transformation?	♦ Which transformations are needed to get from one shape to the other? ♦ What is the least number of moves needed?

5

angle
volume
length

Measurement

perimeter
area

CCSS Measurement: Perimeter

Traditional Formative Assessment

What is the perimeter of the pentagon?

- ○ 9 units
- ○ 13 units
- ○ 15 units
- ○ 18 units

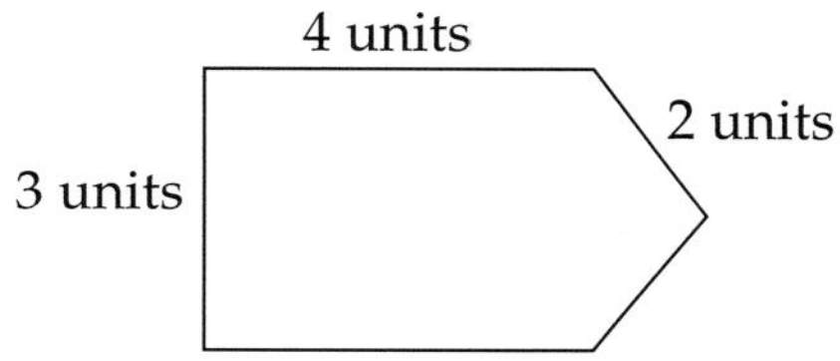

Limitations: Students may know how to find perimeter, but be unable to find the length of the unlabeled sides, or may make a minor computational error and get an incorrect answer.

Enhanced Formative Assessment

Use a ruler to find the perimeter of the pentagon in centimeters.

Eddy

Use a ruler to find the perimeter of the pentagon in centimeters.

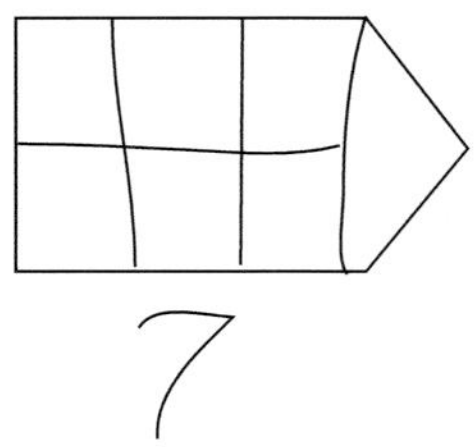

Eddy draws grid lines within the shape and records the total. He confuses perimeter with area.

Instructional Focus: **Rebuild**

Fabayo

Use a ruler to find the perimeter of the pentagon in centimeters.

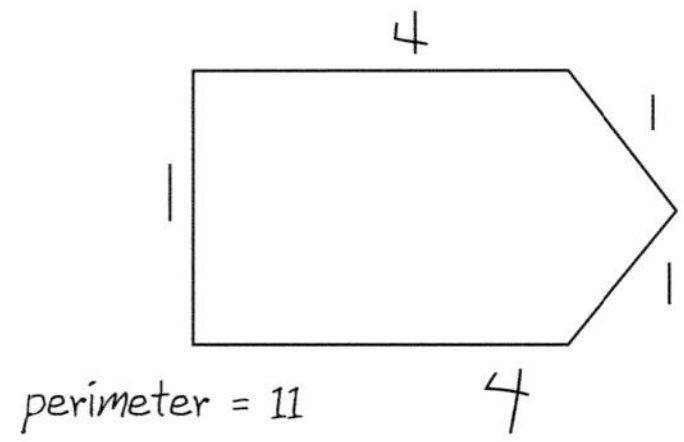

Fabayo measures all sides and calculates the sum. However, some of his measurements are in inches and some are in centimeters.

Instructional Focus: **Core 1**

George

Use a ruler to find the perimeter of the pentagon in centimeters.

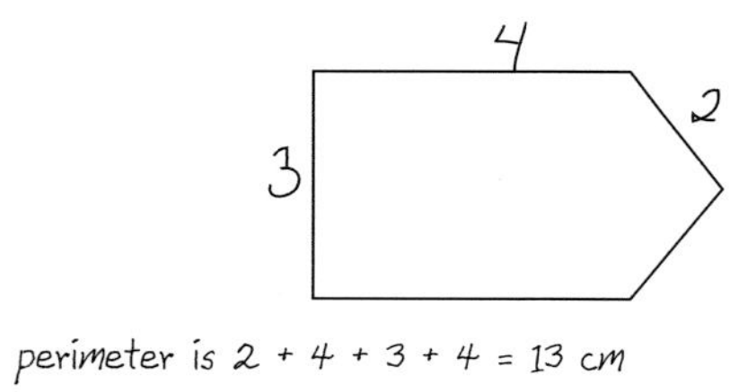

George measures all of the sides of different length. When finding the total, he missed one of the sides.

Instructional Focus: **Core 2**

Helena

Use a ruler to find the perimeter of the pentagon in centimeters.

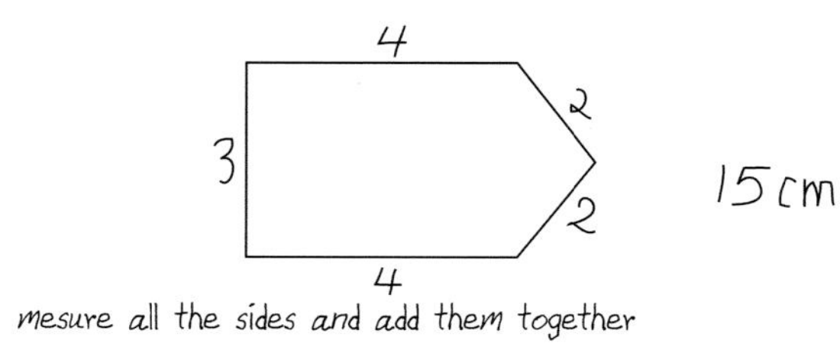

Helena uses the ruler to measure the length of each side in centimeters. She accurately finds the sum of the sides and records the perimeter.

Instructional Focus: **Challenge**

CCSS Measurement: Perimeter

	Rebuild Focus	Core 1 Focus	Core 2 Focus	Challenge Focus
	Bendy Straw Sides	**Draw & Measure**	**√ Check It Out**	**Tangram Creations**
Goal	Define and explore perimeter	Measure length in like units to calculate perimeter	Measure perimeter using a strategy to keep track of sides	Compare perimeters of complex shapes
Materials	♦ Flexible straws ♦ Rulers ♦ Scissors	♦ Rulers (cm and inch) ♦ Drawing paper	♦ Rulers (cm and inch) ♦ Chart paper	♦ Tangram sets ♦ Paper ♦ Scissors ♦ Rulers (cm and inch)
Suggested Activity Directions	Show students how to pinch one end of a straw to insert into a second straw. Allow time for students to cut the flexible straws and use them to make polygons. The straws emphasize the sides of the polygon. Ask students to find the perimeter of the polygons by measuring the length of all sides. Remind students to include the unit when recording perimeter. 	Distribute rulers and allow students time to draw a closed figure. Ask students to compare inches to centimeters. Share that inches are about 2½ times as long as a centimeter. Remind students to use the same unit when measuring sides to find perimeter. As students measure each side, challenge them to estimate the length based on the sides already measured to get an idea of an approximate measure. 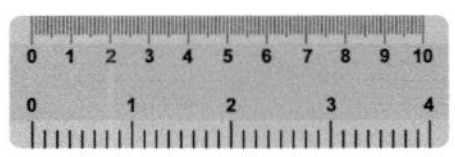	Provide time for students to draw multisided polygons on the chart paper. Discuss how to measure and find the perimeter. Ask students how to keep track of sides already measured. As a group, generate a few ideas (e.g., check mark the length; use a dash). Invite pairs of students to use one of the strategies when finding perimeter of the shapes. 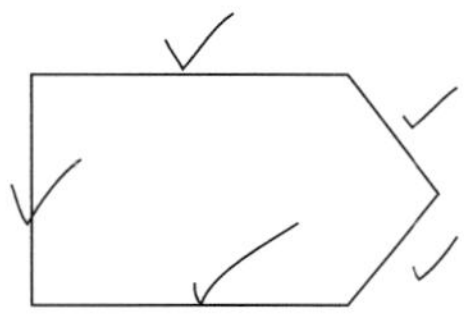	Provide students with tangrams. The set includes 7 tangrams: 2 large triangles, 1 medium triangle, 2 small triangles, 1 parallelogram, and 1 square. Provide time for students to make a closed figure using all 7 tangrams. Ask students to cut out the new shape and measure the perimeter. Discuss how the perimeters of the new shapes differ from one another even though they are made from the same 7 pieces. 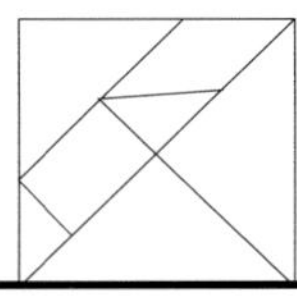
Questions to Assess	♦ How many sides does your polygon have? ♦ What is the sum of all of the lengths?	♦ How do inches compare to centimeters? ♦ If the short side was 4 centimeters, about how long will this side be?	♦ How can we keep track of sides already measured? ♦ What strategy can you use to check your measurements and calculations?	♦ If the areas of the figures are the same will the perimeter be the same? ♦ What do you notice when you order the figures according to perimeter?

Traditional Formative Assessment

What is the area of the rectangle?

Limitations: If a student answers incorrectly, it might be difficult to determine if an error was made in defining area, measuring the sides, or calculating.

Enhanced Formative Assessment

Explain two different ways to find the area of a rectangle. Use words, pictures, or numbers.

CCSS Measurement: Area

Ivan

Explain two different ways to find the area of a rectangle. Use words, pictures, or numbers.

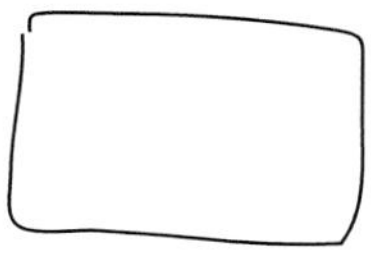

Ivan simply draws a rectangle with no mention of area.

Instructional Focus: **Rebuild**

Joyce

Explain two different ways to find the area of a rectangle. Use words, pictures, or numbers.

It's the inside.

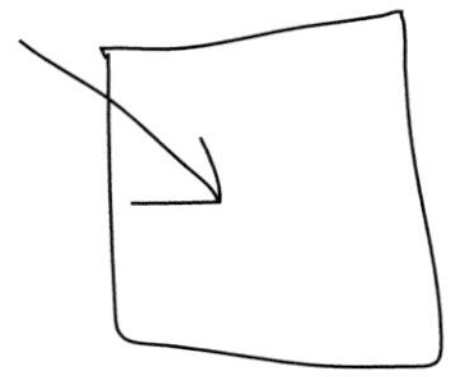

Joyce shows that area is found within the shape, but does not offer a strategy to find area.

Instructional Focus: **Core 1**

Karl

Explain two different ways to find the area of a rectangle. Use words, pictures, or numbers.

Draw the rectangle on a grid and count the squares inside.

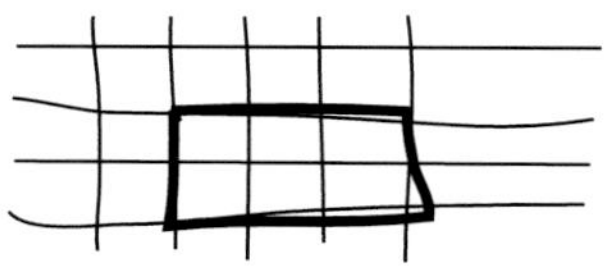

area = 6 sqar units

Karl shows one correct strategy to find the area of a rectangle.

Instructional Focus: **Core 2**

Lana

Explain two different ways to find the area of a rectangle. Use words, pictures, or numbers.

Find the length and width and multiply to get area. Cover with tiles and count all tiles.

Lana accurately describes 2 strategies to find the area of a rectangle.

Instructional Focus: **Challenge**

CCSS Measurement: Area

<table>
<tr><th></th><th>Rebuild Focus</th><th>Core 1 Focus</th><th>Core 2 Focus</th><th>Challenge Focus</th></tr>
<tr><th></th><th>Color & Compare</th><th>Fill It Up!</th><th>Arrays & Area</th><th>The Area Is...</th></tr>
<tr><th>Goal</th><td>Define and compare area</td><td>Measure area in standard units</td><td>Measure area of rectangles using length times width</td><td>Measure area of other polygons</td></tr>
<tr><th>Materials</th><td>♦ Rulers
♦ Paper
♦ Crayons
♦ Scissors</td><td>♦ Tiles (1 inch)
♦ Grid paper (1 inch)</td><td>♦ Multiplication chart (see page 107)
♦ Transparencies</td><td>♦ Markers
♦ Paper triangles and trapezoids
♦ Scissors</td></tr>
<tr><th>Suggested Activity Directions</th><td>Ask students to draw and cut out various shapes. Explain area as the amount of space a figure covers. Allow students to color the shapes and compare how much space the shape covers. Ask students to compare areas of shapes by overlapping the shapes and describing the size. Challenge students to sequence all the shapes according to area and prove the order is correct.
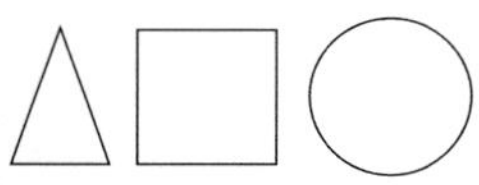</td><td>Distribute 12 tiles to each student. Ask students to make several closed figures using all 12 tiles. Model for students how to outline the figure on the grid paper and record the area in square units. Allow students to make additional shapes with various areas. Look for noncongruent shapes with the same area.
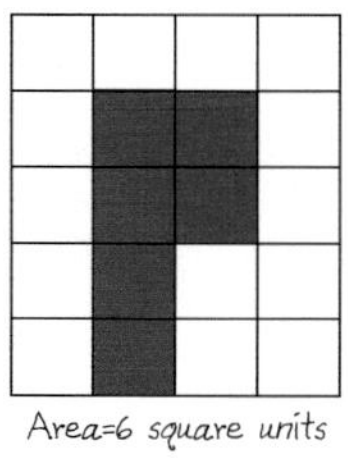
Area=6 square units</td><td>Model for students how to draw a rectangle on the multiplication chart beginning at the top, left corner. Ask students to record length, width, and area on a chart. Ask students to describe the connections they see.
<table>
<tr><th>x</th><th>1</th><th>2</th><th>3</th><th>4</th><th>5</th><th>6</th><th>7</th><th>8</th><th>9</th></tr>
<tr><th>1</th><td>1</td><td>2</td><td>3</td><td>4</td><td>5</td><td>6</td><td>7</td><td>8</td><td>9</td></tr>
<tr><th>2</th><td>2</td><td>4</td><td>6</td><td>8</td><td>10</td><td>12</td><td>14</td><td>16</td><td>18</td></tr>
<tr><th>3</th><td>3</td><td>6</td><td>9</td><td>12</td><td>15</td><td>18</td><td>21</td><td>24</td><td>27</td></tr>
<tr><th>4</th><td>4</td><td>8</td><td>12</td><td>16</td><td>20</td><td>24</td><td>28</td><td>32</td><td>36</td></tr>
<tr><th>5</th><td>5</td><td>10</td><td>15</td><td>20</td><td>25</td><td>30</td><td>35</td><td>40</td><td>45</td></tr>
<tr><th>6</th><td>6</td><td>12</td><td>18</td><td>24</td><td>30</td><td>36</td><td>42</td><td>48</td><td>54</td></tr>
<tr><th>7</th><td>7</td><td>14</td><td>21</td><td>28</td><td>35</td><td>42</td><td>49</td><td>56</td><td>63</td></tr>
<tr><th>8</th><td>8</td><td>16</td><td>24</td><td>32</td><td>40</td><td>48</td><td>56</td><td>64</td><td>72</td></tr>
<tr><th>9</th><td>9</td><td>18</td><td>27</td><td>36</td><td>45</td><td>54</td><td>63</td><td>72</td><td>81</td></tr>
</table>
<table>
<tr><th>l</th><th>w</th><th>area</th></tr>
<tr><td>3</td><td>2</td><td>6 sq. units</td></tr>
<tr><td>4</td><td>3</td><td>12 sq. units</td></tr>
<tr><td>6</td><td>3</td><td>18 sq. units</td></tr>
<tr><td></td><td></td><td></td></tr>
</table></td><td>Challenge students to apply their knowledge of finding the area of a rectangle to find the area of triangles or trapezoids. Encourage students to fold, cut, or manipulate the shapes as they try different strategies. Encourage students to share strategies with the group. Discuss ways to represent the strategies symbolically.
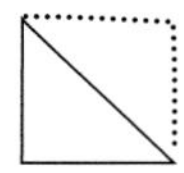
Area of right triangle = l x w ÷ 2</td></tr>
<tr><th>Questions to Assess</th><td>♦ Which shape covers the greatest area?
♦ How could you prove that one area is less than another?</td><td>♦ What is the name of the shape?
♦ Can you make two noncongruent hexagons with the same area?</td><td>♦ What relationship do you notice between the length, width, and area?
♦ How does the multiplication chart help you find the area of the rectangles?</td><td>♦ How does the area of the shape relate to the area of a rectangle?
♦ What shapes do you see within the trapezoid?</td></tr>
</table>

CCSS Measurement: Volume

Traditional Formative Assessment

What is the volume of a cube if one side is 3 cm?

- O 3 units
- O 9 square units
- O 9 cubic units
- O 27 square units
- O 27 cubic units

Limitations: An incorrect answer will be difficult to diagnose. A visual of a cube would help students focus on the concept of volume rather than wondering if the student was simply unable to visualize a cube.

Enhanced Formative Assessment

What is the volume of the cube? Show your work.

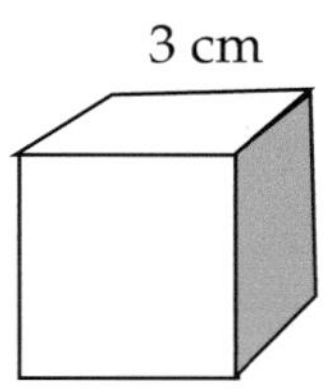

Note: Figure not drawn to scale.

CCSS Measurement: Volume

Melissa

What is the volume of the cube? Show your work.

3 cm

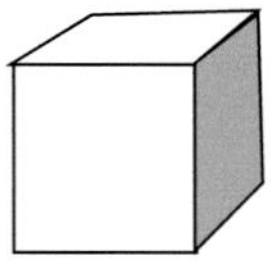

loud

Melissa does not recognize “volume” as a mathematics term.

Instructional Focus: **Rebuild**

Neil

What is the volume of the cube? Show your work.

3 cm

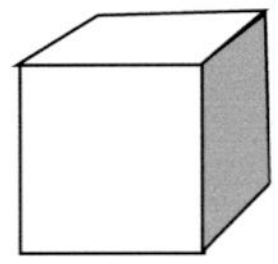

$3 \times 3 = 9$

Neil shows how to find the area of 1 square face rather than the volume of the cube.

Instructional Focus: **Core 1**

Odell

What is the volume of the cube? Show your work.

3 cm

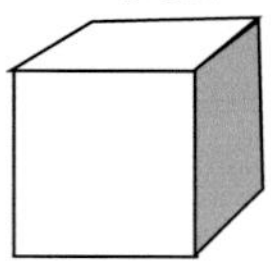

I cud use cm cubes to make one the same size and count the cubes. It wud be 3 tall, 3 wide & 3 long.

Odell describes how to rebuild the cube so she can count the cubes to find the volume.

Instructional Focus: **Core 2**

Phylla

What is the volume of the cube? Show your work.

3 cm

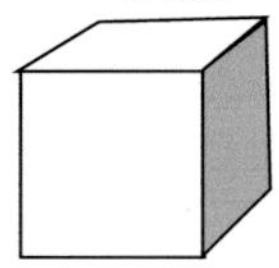

length x width x hight

$3 + 3 + 3 = 27 cm^3$

Phylla correctly finds the volume and labels her answer with the appropriate unit.

Instructional Focus: **Challenge**

CCSS Measurement: Volume

<table>
<tr><th></th><th>Rebuild Focus</th><th>Core 1 Focus</th><th>Core 2 Focus</th><th>Challenge Focus</th></tr>
<tr><th></th><th>Build & Count</th><th>How Many?</th><th>What's the Formula?</th><th>What If?</th></tr>
<tr><th>Goal</th><td>Define and compare volume</td><td>Measure volume in cubic centimeters</td><td>Measure volume of rectangular prisms using length times width times height</td><td>Measure volume when dimensions change</td></tr>
<tr><th>Materials</th><td>♦ Multilink cubes
♦ Rectangular prisms</td><td>♦ Centimeter cubes</td><td>♦ Centimeter cubes
♦ Variety of boxes
♦ Calculators</td><td>♦ Centimeter grid paper
♦ Centimeter cubes
♦ Calculators</td></tr>
<tr><th>Suggested Activity Directions</th><td>Model how to build rectangular prisms using the multilink cubes. Define volume as the size of solid objects. Ask students to put the prisms in order from greatest to smallest volume. Encourage students to use efficient counting strategies (e.g., skip counting, counting by layers) to compare the volume of the prisms.
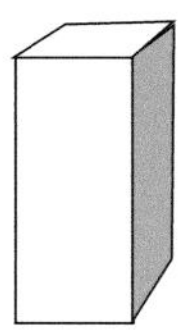</td><td>Model building a rectangular prism using the centimeter cubes. Challenge students to make as many different rectangular prisms as possible using exactly 12 centimeter cubes. Allow students to describe the differences in dimensions among the prisms with the same volume. Try again using 24 cubes or 36 cubes. Remind students to state the volume in cubic centimeters.</td><td>Choose a box and ask students to find the volume in cubic centimeters. Choose a box large enough there are not enough cubes to fill the box. Challenge students to use the cubes they have to develop a strategy. Model recording the attribute and dimensions. Allow students to use calculators to compute volume.
<table>
<tr><th></th><th>Box 1</th><th>Box 2</th></tr>
<tr><td>Length</td><td>10 cm</td><td>12 cm</td></tr>
<tr><td>Width</td><td>3 cm</td><td>3 cm</td></tr>
<tr><td>Height</td><td>2 cm</td><td>4 cm</td></tr>
<tr><td></td><td></td><td></td></tr>
<tr><td>Volume</td><td>60 cm^3</td><td>144 cm^3</td></tr>
</table></td><td>Pose the following questions for students to explore:
♦ If the length of a rectangular prism was cut in half, how would it affect the volume?
♦ If the width of a rectangular prism is doubled, how would it affect the volume?
♦ If all three dimensions were cut in half, how would it affect the volume?
</td></tr>
<tr><th>Questions to Assess</th><td>♦ How do you know the prism has a greater volume?
♦ What counting strategy could we use?</td><td>♦ If the volume is the same, what is different among the shapes?
♦ What are the three dimensions of the rectangular prism?</td><td>♦ What strategy did you use when you did not have enough centimeter cubes to fill the box?
♦ Would the strategy work without the blocks?</td><td>♦ If the volume is cut in half, what could the dimensions be?
♦ How are the cubic units affected by a single dimension?</td></tr>
</table>

Traditional Formative Assessment

The angle measures __________ degrees.

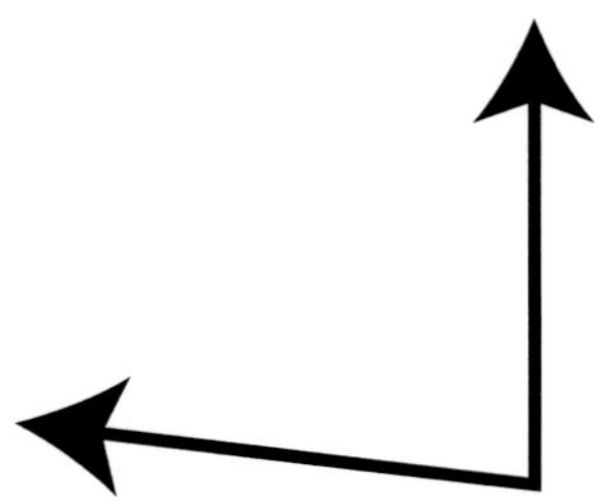

Limitations: Students may be unable to use a protractor but still have conceptual knowledge of angle measurements, which goes untapped in this assessment item.

Enhanced Formative Assessment

Describe the measurement of the angle.

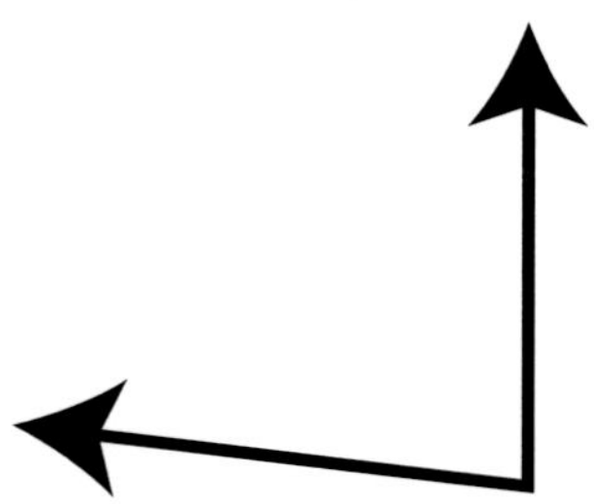

CCSS Measurement: Angles

Quenton

Describe the measurement of the angle.

I made a triangle.

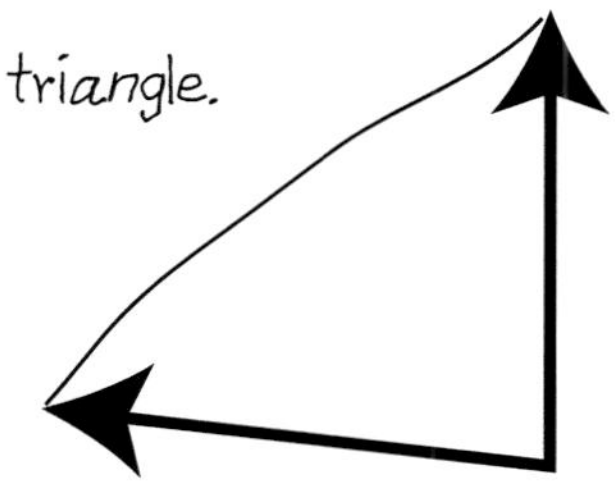

Quenton joins the rays and makes a triangle, which contains the word "angle."

Instructional Focus: **Rebuild**

Sandy

Describe the measurement of the angle.

Left angle

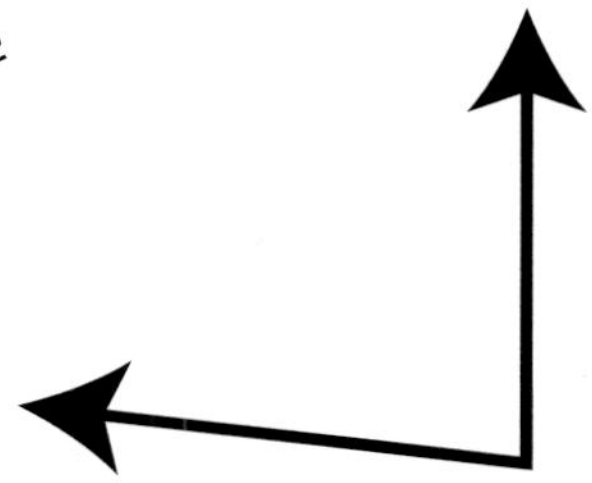

Sandy calls the angle a left angle because the orientation is the opposite of the examples of right angles she has seen.

Instructional Focus: **Core 1**

Tracey

Describe the measurement of the angle.

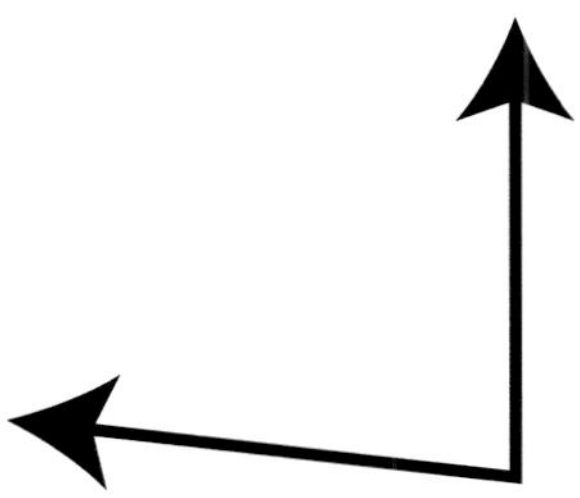

The angle is a little smaller than a right angle. It is close to 90 degres.

Tracey uses her knowledge of right angles to describe the angle as close to 90 degrees.

Instructional Focus: **Core 2**

Ulysses

Describe the measurement of the angle.

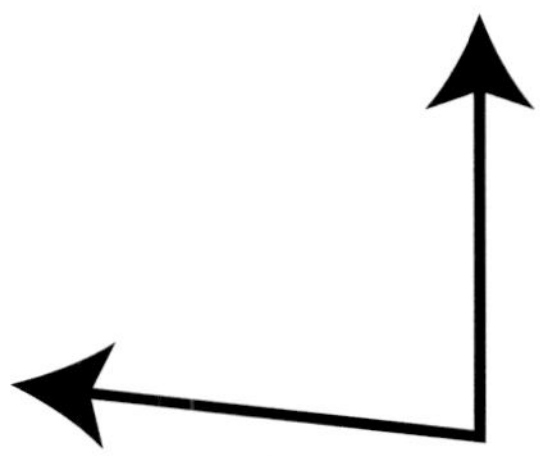

The angle inside is 85°
so the outside angle is 275°.
Together it is 360° like in a circle.

Ulysses correctly measures the angle and describes the exterior angle measure based on the interior angle.

Instructional Focus: **Challenge**

CCSS Measurement: Angles

	Rebuild Focus **Bigger or Smaller**	Core 1 Focus **Angle Tangle**	Core 2 Focus **Making Protractors**	Challenge Focus **Making Spinners**
Goal	Compare angles	Classify angles	Measure angles	Calculate angle measurements in a spinner
Materials	♦ Tag board strips ♦ Brass fastener ♦ Paper	♦ Pattern blocks	♦ Circle cutouts ♦ Protractors ♦ Scissors	♦ Compass ♦ Protractor ♦ Rulers ♦ Calculators
Suggested Activity Directions	Fasten 2 tag board strips with a brass fastener. Ask students to locate angles around the room and adjust their angle tools to the same measure. Demonstrate for students how to trace the angle on a sheet of paper. Provide an opportunity for students to draw several angles and then number the angles according to size. Students may use their angle tool to confirm the comparison. 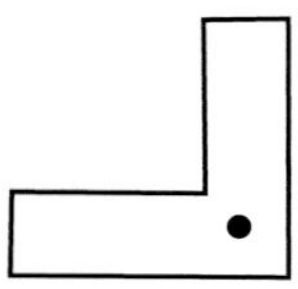	Allow students time to explore the pattern blocks. Model use of the orange square to name angles larger or smaller than 90°. Find the angle measures of all of the angles on the pattern blocks by only using the pattern blocks. Invite students to share their strategies for determining the measurement of each angle. Use the pattern blocks to measure additional angles throughout the classroom. 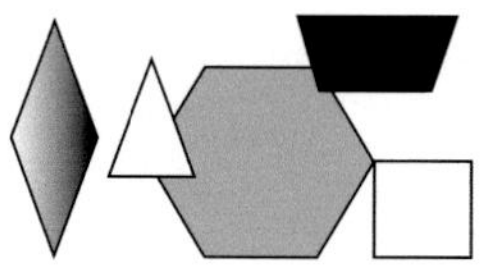	Explain to students that a circle has 360°. Fold the circle in half and ask the angle measurement. Continue folding and calculating the angle measurement until the circle is divided into eight sections. Model how to cut a 45°, 90°, and 225° angle. Use combinations of the pieces to the measure angles around the room. Let students explore with the protractor and discuss how the protractor they made compares to the commercially made protractors. 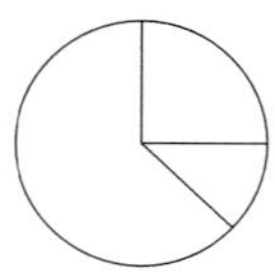	Remind students how frequently spinners are used in math. Explain that they will make spinners to use in the classroom. Model using the compass to draw circles. Ask students to make spinners with 4, 5, 6, and 8 equal sections. Encourage students to discuss strategies among their peers. Students may develop their own math games to play using the spinners they created. 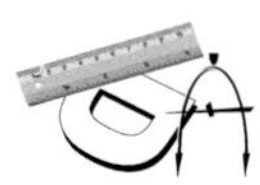
Questions to Assess	♦ Which angle is the smallest? ♦ Does the angle make a square corner?	♦ Are the angles acute, obtuse, or right angles? ♦ Which angles are the same?	♦ What is a reasonable range for the angle measurement? ♦ Is the angle larger or smaller than 180°?	♦ How many sections are needed for the spinner? ♦ What is the angle measure of each section?

6

Data Analysis & Probability

circle leaf bar line graphs stem

Traditional Formative Assessment

The graph shows the number of activities each student completed on Field Day.

How many activities did Sonya complete?

- O 14
- O 16
- O 17
- O 18

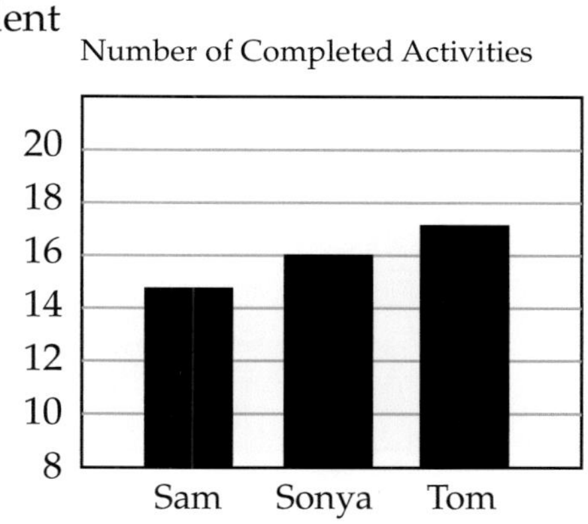

Limitations: The question has a low level of cognitive demand and does not assess knowledge of constructing or interpreting bar graphs. Also, the graph is truncated.

Enhanced Formative Assessment

Use the data from the table to make a bar graph.
Include all the parts of a bar graph.

Field Day Activities Completed	
Name	**Number**
Sam	14
Sonya	15
Tom	16
Leo	13

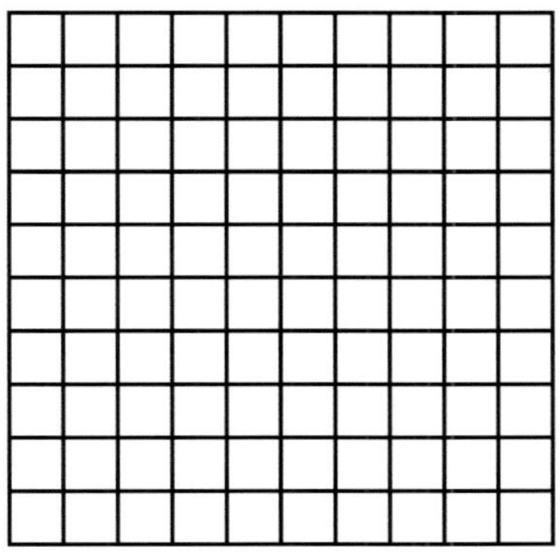

Aasha

Use the data from the table to make a bar graph.
Include all the parts of a bar graph.

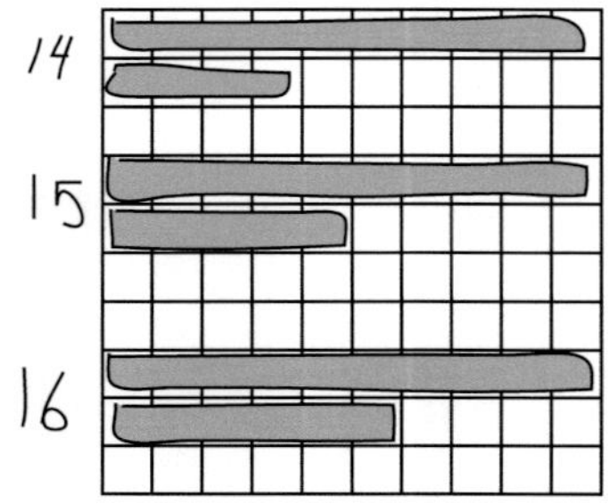

Aasha shows the values 14, 15, and 16 from the table, but not in a common bar graph format.

Instructional Focus: **Rebuild**

Balin

Use the data from the table to make a bar graph.
Include all the parts of a bar graph.

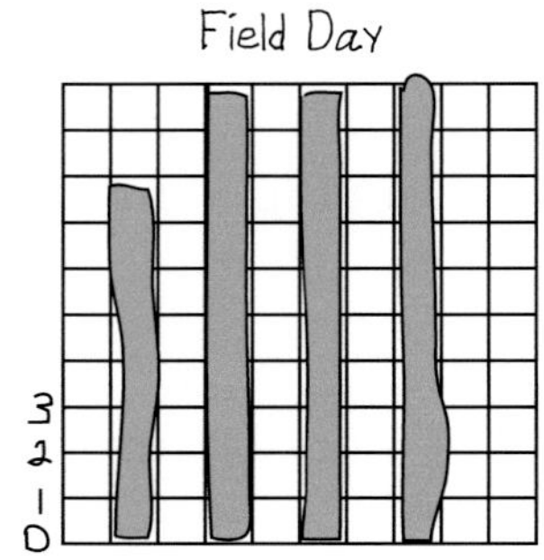

Graph is not tall enuf to make.
It only goes to 9.

Balin says there are not enough spaces to available to finish the graph. He cannot determine an appropriate scale based on the data.

Instructional Focus: **Core 1**

Cara

Use the data from the table to make a bar graph.
Include all the parts of a bar graph.

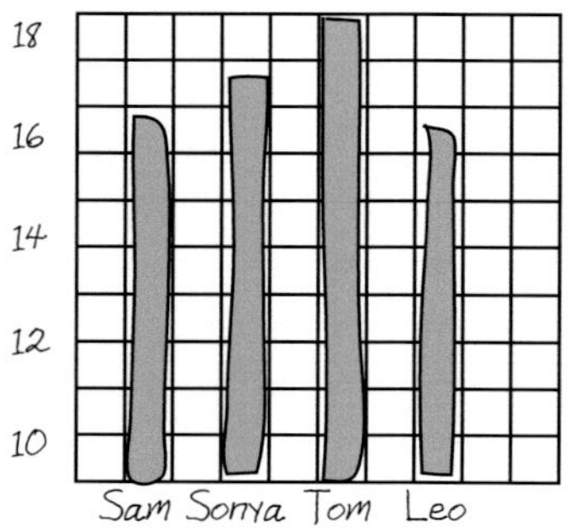

Cara correctly shades the graph and uses an appropriate scale, but does not include a title and labels.

Instructional Focus: **Core 2**

Daryl

Use the data from the table to make a bar graph.
Include all the parts of a bar graph.

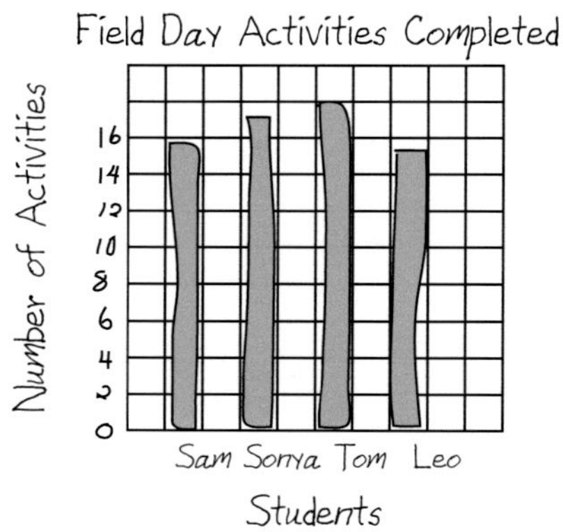

Daryl includes an appropriate scale, labels, and a title. He graphs the data correctly.

Instructional Focus: **Challenge**

Data Analysis & Probability: Bar Graphs

<table>
<tr><th></th><th>Rebuild Focus
For Keeps</th><th>Core 1 Focus
What Works?</th><th>Core 2 Focus
Cut & Paste</th><th>Challenge Focus
Double Data</th></tr>
<tr><td>Goal</td><td>Represent data from a real graph on a bar graph</td><td>Choose an appropriate scale given the data set</td><td>Use titles from a table on a bar graph</td><td>Create double bar graphs</td></tr>
<tr><td>Materials</td><td>♦ Large graph paper
♦ Graph paper
♦ Toy cars (or other tokens)</td><td>♦ Bar graph-enhanced formative assessment
♦ Graph paper</td><td>♦ Bar graph-enhanced formative assessment
♦ Scissors
♦ Glue sticks</td><td>♦ Graph paper</td></tr>
<tr><td>Suggested Activity Directions</td><td>Ask students to sort a collection of cars and name the groups (e.g., color, style, size). Allow students to decide which attribute they would like to graph. Decide what the categorical labels will be and give students an opportunity to place the cars on the paper to make a real graph. Explain the need to record the data so the information is not lost when the cars are put away. As each car is removed, shade the space. Add labels, a scale, and a title. Pose questions about the data.</td><td>Lead students in a skip-counting exercise. Practice skip-counting by 2, 3, 4, 5, 10, and 20. Challenge students to discuss how to display data when there are a limited number of spaces. Explore different scales and check to see if they fit given the data values. Allow students to display the same values on graphs with different scales. Compare the appearance of the graphs.

3, 6, 9, 12,...</td><td>Allow time for students to represent the data on the bar graph by shading and choosing an appropriate scale. Discuss examples of different scales which would work. Ask students to cut the title and the headings from the table and place them in the correct location on the bar graph. Prompt students to note the connection between the two displays.

<table><tr><th colspan="2">Field Day Activities Completed</th></tr><tr><th>Name</th><th>Number</th></tr></table></td><td>Create a double bar graph using the data from the table below. Model for students how to show double bars for each name to represent the number of A.M. and P.M. activities. Ask students to share observations about the data. Discuss other data which may be best displayed as a double bar graph rather than a single bar graph or two separate bar graphs.
<table><tr><th colspan="3">Field Day Activities</th></tr><tr><th>Name</th><th>PM</th><th>AM</th></tr><tr><td>Sam</td><td>8</td><td>6</td></tr><tr><td>Sonya</td><td>10</td><td>5</td></tr><tr><td>Tom</td><td>8</td><td>8</td></tr><tr><td>Leo</td><td>9</td><td>4</td></tr></table></td></tr>
<tr><td>Questions to Assess</td><td>♦ How can each car be represented on the graph?
♦ How will others know what the graph is showing?</td><td>♦ What is the largest value in the data set?
♦ If we use a scale of 5, what is the greatest value that will fit on the graph paper?</td><td>♦ What do the labels communicate about the data?
♦ What is the purpose of a title?</td><td>♦ Why might these data be collected?
♦ What other data sources might be displayed on a double bar graph?</td></tr>
</table>

Traditional Formative Assessment

What is the range of the number of foul shots?

- O 3
- O 4
- O 6
- O 7

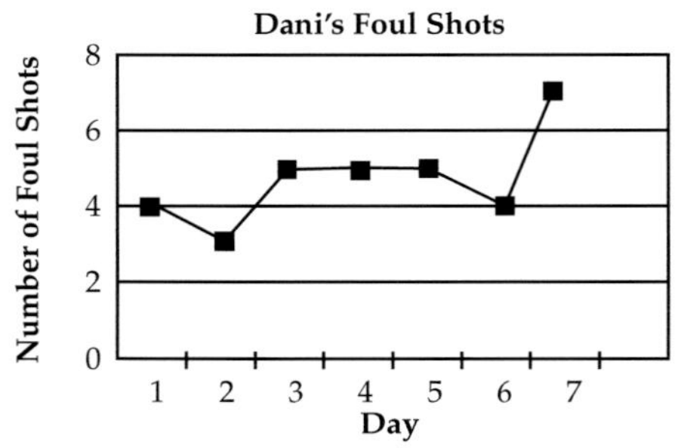

Limitations: An incorrect response may be due to a lack of understanding about range rather than an ability to read and interpret the line graph.

Enhanced Formative Assessment

The table shows how many foul shots Dani made each day.
User the table to make a line graph.

Day	1	2	3	4	5	6	7
Shots Made	4	2	3	5	5	5	7

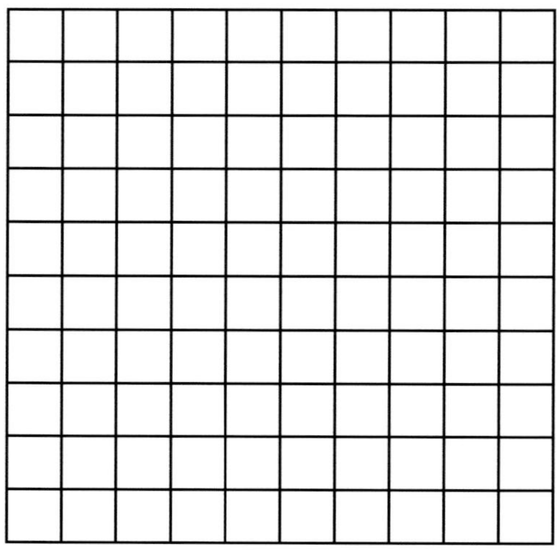

Data Analysis & Probability: Line Graphs

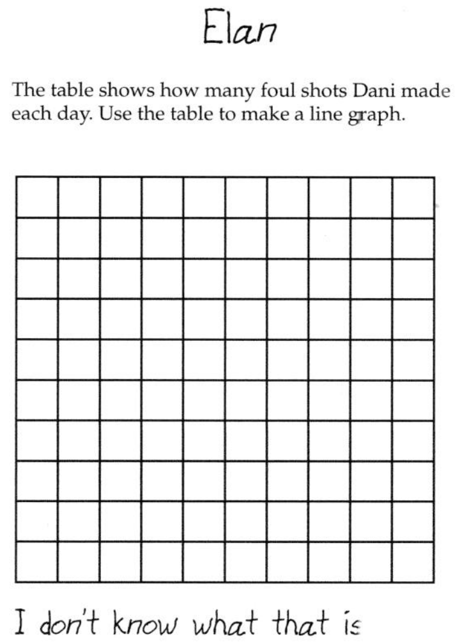

Elan does not recognize the term line graph and makes no attempt to create a graph.
Instructional Focus: **Rebuild**

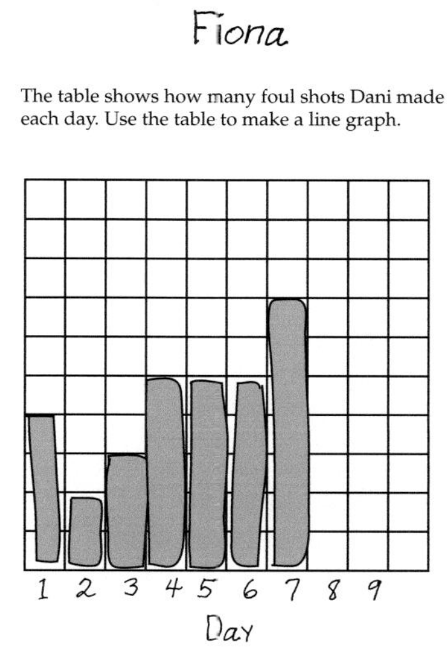

Fiona makes a data display similar to a bar graph.
Instructional Focus: **Core 1**

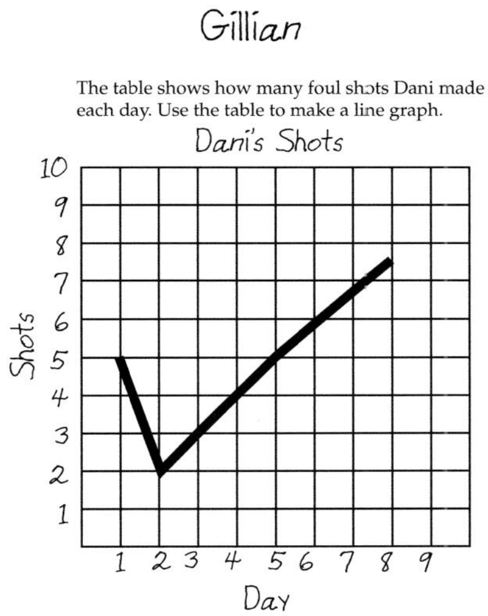

Gillian has all of the key elements of a line graph but incorrectly plots some of the points.
Instructional Focus: **Core 2**

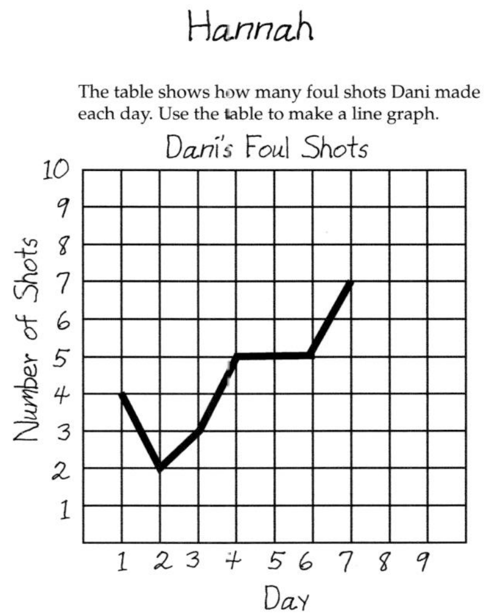

Hannah correctly displays the data on a line graph and includes a scale, labels, and a title.
Instructional Focus: **Challenge**

Data Analysis & Probability: Line Graphs

	Rebuild Focus	Core 1 Focus	Core 2 Focus	Challenge Focus
	Change Over Time	**Lines or Bars**	**Shoot & Plot**	**Seeing Double**
Goal	Understand the elements of a line graph	Understand that a line graph shows change over time	Plot points on a line graph	Create double-line graphs
Materials	♦ Graph paper ♦ Counters ♦ Paper ball ♦ Trashcan	♦ Graph paper ♦ Counters ♦ Paper ball ♦ Trashcan	♦ Graph paper ♦ Counters ♦ Paper ball ♦ Trashcan	♦ Graph paper ♦ Markers ♦ Paper ball ♦ Trashcan
Suggested Activity Directions	Prepare a line graph using data collected while students take 10 "foul shots" in each of 5 rounds using paper balls and trashcan. Explain that we want to see a picture of the data so we can find out if our performance changed over time. Model for students how to plot the points. Talk with students to generate correct labels and a title. Ask students to describe the line on the graph. Discuss how the line changes from round to round as time passes. 	Discuss with students that a line graph shows change over time whereas a bar graph often shows categorical data. Discuss examples of each. Prepare a line graph with students as they take 10 foul shots over 5 rounds. Record the results of each round. Guide students in the construction of the line graph, connecting the plotted points to show change over time. 	Plot points on a line graph as students take foul shots and record the total baskets made for each of 5 rounds. Provide graph paper and record the number of rounds on the x-axis. The y-axis will indicate the number of baskets made. Allow students to use a ball of paper and the trashcan to take 10 foul shots. The student then places the counter on the line graph at the point of intersection. A partner checks the point for accuracy, and if agrees, the point is plotted.	Prepare a line graph using data collected while taking foul shots. Allow students to use a ball of paper and the trashcan to take foul shots. On turn, each student stands an equal distance from the trashcan and takes 10 shoots. Model for students how to record the data on a table. Students each take four turns. Ask pairs of students to display the data from the table on one line graph. Each player's line should be a different color. 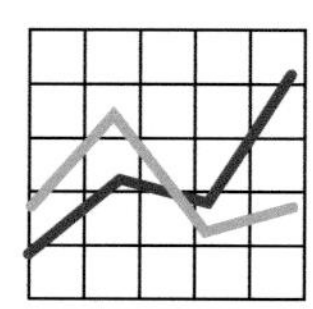
Questions to Assess	♦ How many baskets did you make? ♦ How does the direction of the line change with the data?	♦ What are some things that change over time? ♦ What are some examples of categorical data?	♦ How do you know the point is in the correct place? ♦ Do your data show a trend? If so, what is the trend after 5 rounds?	♦ How do your data compare to your partner's data? ♦ Did your performance improve? How do you know?

Traditional Formative Assessment

The line plot shows the length of twenty-five boats in the bay.

Boats in the Bay

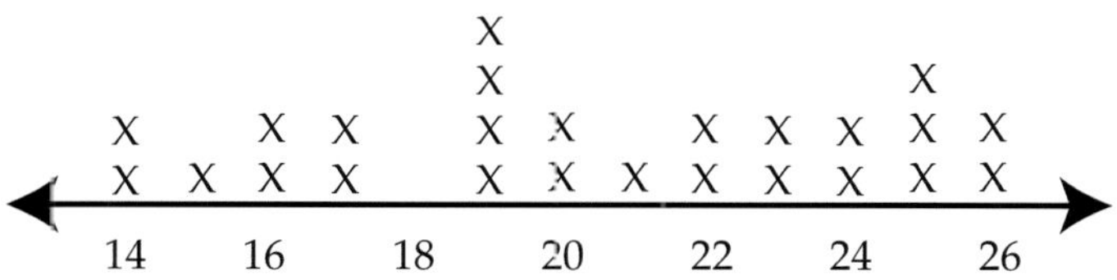

How many of the boats are 19 feet long?

- O 0
- O 2
- O 3
- O 4

Limitations: The number line should include tick marks to provide a visual for students to represent the values on the number line as only the even values are listed.

Enhanced Formative Assessment

The line plot shows the distance traveled by twelve cyclists. Make a list of the distances displayed on the line plot and write one fact about the line plot.

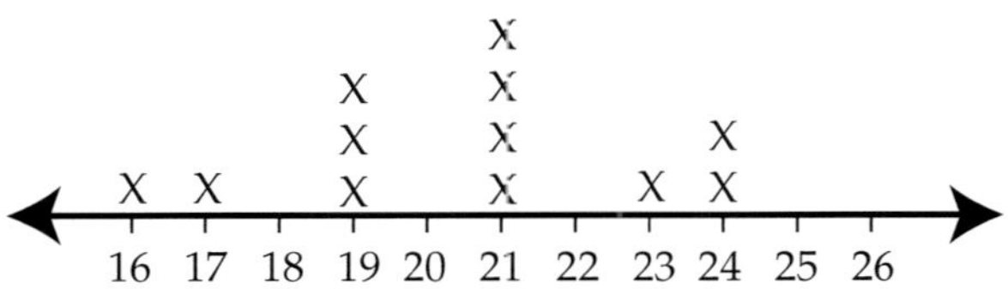

John

The line plot shows the distance traveled by twelve cyclists. Make a list of the distances displayed on the line plot and write one fact about the line plot.

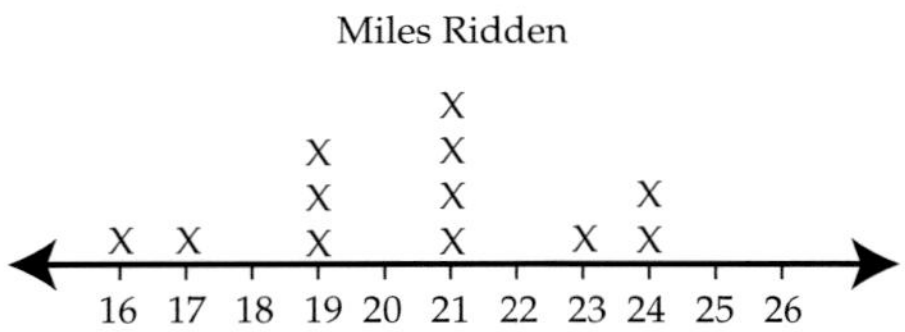

12 XS

John counts the number of Xs on the line plot and does not provide any relevant facts.

Instructional Focus: **Rebuild**

Kym

The line plot shows the distance traveled by twelve cyclists. Make a list of the distances displayed on the line plot and write one fact about the line plot.

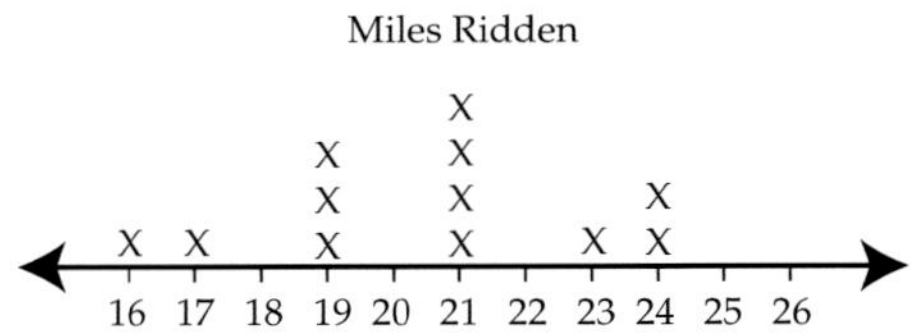

16, 17, 18, 19, 20, 21, 22, 23, 24, 25, 26

Kym lists all of the values on the number line rather than the distance of each cyclist.

Instructional Focus: **Core 1**

Louise

The line plot shows the distance traveled by twelve cyclists. Make a list of the distances displayed on the line plot and write one fact about the line plot.

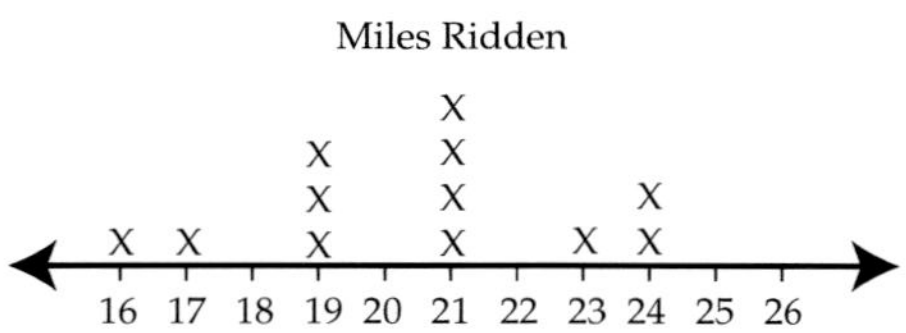

3 peple went 19 miles.

Louise provides a true fact, but does not list the distances of each cyclist.

Instructional Focus: **Core 2**

Mike

The line plot shows the distance traveled by twelve cyclists. Make a list of the distances displayed on the line plot and write one fact about the line plot.

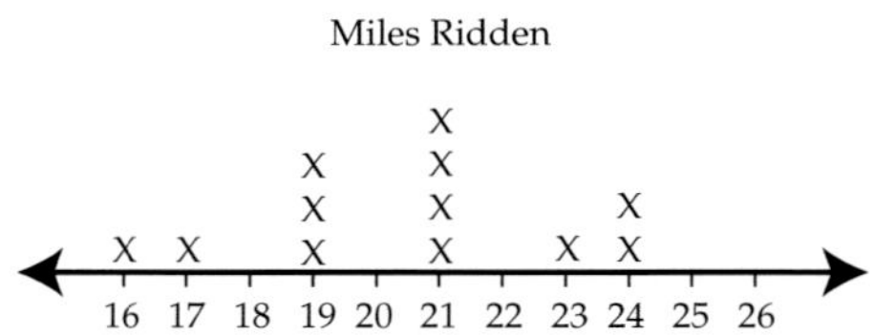

16, 17, 19, 19, 19, 21, 21, 21, 21, 23, 24, 24
More bikers went 21 miles than any other distance and nobody went 18 miles.

Mike lists the correct data set and offers two facts about the line plot.

Instructional Focus: **Challenge**

CCSS Data Analysis & Probability: Line Plots

	Rebuild Focus	Core 1 Focus	Core 2 Focus	Challenge Focus
	That's Mine	**I Want to Know**	**X Marks the Spot**	**Tower Time**
Goal	Construct line plots with manipulatives	Describe line plots	Identify the data values on a line plot	Find measures of central tendency
Materials	♦ Dry erase boards ♦ Markers ♦ Unifix cubes	♦ Dry erase boards ♦ Markers ♦ Unifix cubes	♦ Dry erase boards ♦ Markers ♦ Unifix cubes ♦ Sticky notes	♦ Dry erase boards ♦ Markers ♦ Unifix cubes ♦ Sticky notes
Suggested Activity Directions	Create a line plot using cubes to represent the data points. Display data of personal interest to students such as age, shoe size, or number of family members. Model drawing the number line. Ask each student the survey question. Students add a cube to the line plot to represent responses to the survey question. Plot the data points one at a time. Share observations about the line plot. 	Explain that line plots can be used to collect data as well as display data. Share with students that line plots always include a number line, so the data are quantities rather than a categories. Brainstorm a few survey questions student could ask. Discuss the possible range of the number line. Provide time for students to survey their peers and record the data on the dry erase board in a line plot. Ask students to share their results.	Give students 1 minute to build a tower of cubes. Ask students to record their tower quantity on a sticky note. Ask students to organize the sticky notes to form a line plot. Discuss the values needed on the number line to include all of the values. Allow students to draw the number line and represent the values on the sticky notes with Xs on the line plot. Remove the sticky notes one at a time as the Xs are added to the line plot.	Give students one minute to connect as many cubes as possible. Ask each student to record their quantity on a sticky note. Discuss ways to organize the sticky notes. Provide time for students to construct line plots on the dry erase board to represent the cube quantities. Discuss the mode, median, and mean of the data displayed on the line plot. Describe strategies used to find each measurement. 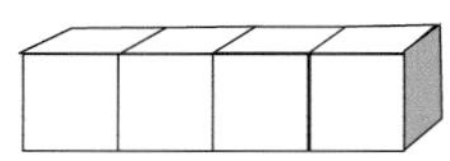
Questions to Assess	♦ How many people are older than you? ♦ Most families are how large?	♦ How many people did you survey? ♦ How did you record each response?	♦ What does the X represent? ♦ How do I know how many data values are included on the line plot?	♦ How do you find the median? ♦ How do the mean, median, and mode compare to one another?

Data Analysis & Probability: Likelihood of an Event

Traditional Formative Assessment

Look at the set of shapes. What is the probability of picking a rectangle?

O 1/4

O 2/5

O 3/5

O 4/6

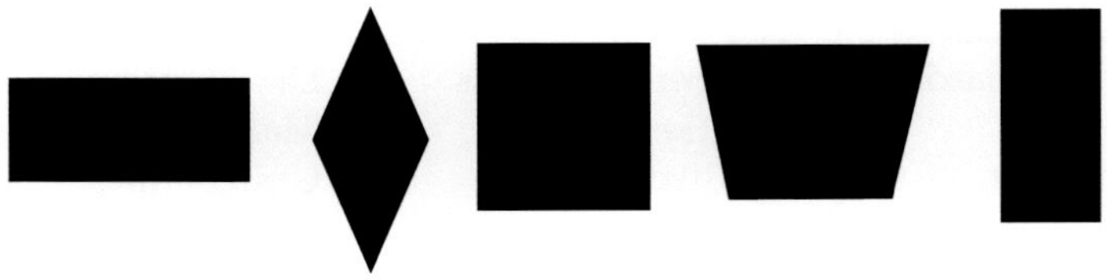

Limitations: In addition to the common problems with multiple choice questions, further confusion may occur as students decide whether or not to include a square as a rectangle.

Enhanced Formative Assessment

What is the probability of spinning a four-sided shape?
Explain your answer.

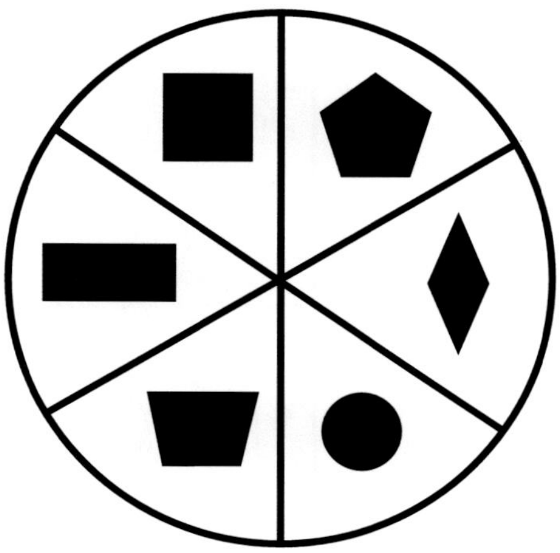

Data Analysis & Probability: Likelihood of an Event

Safara

What is the probability of spinning a four-sided shape? Explain your answer.

6 becuse there are 6 shapes

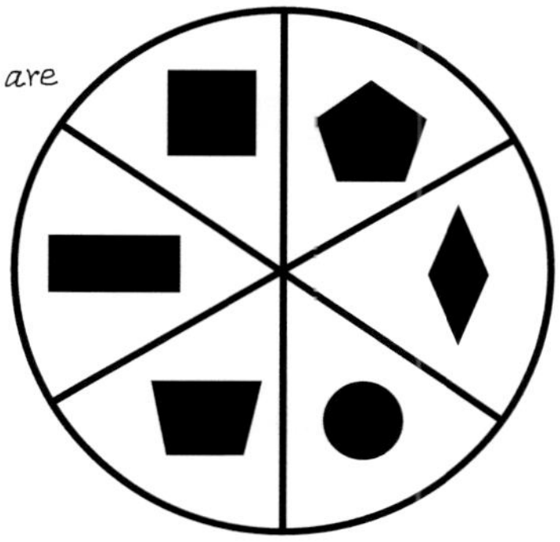

Safara counts the shapes and record the total. She does not record the probability.

Instructional Focus: **Rebuild**

Tamir

What is the probability of spinning a four-sided shape? Explain your answer.

More likely than a circle.

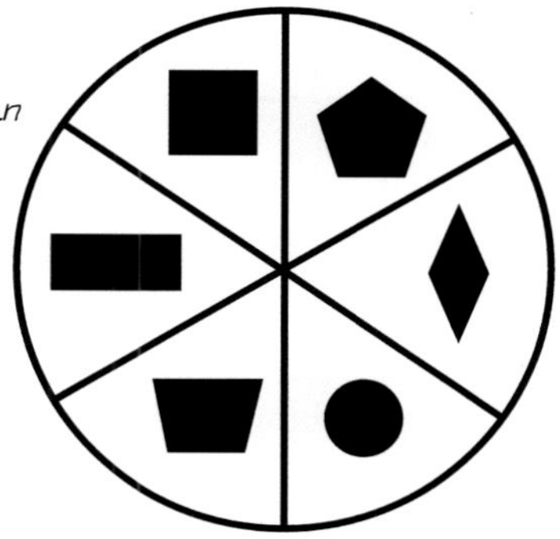

Tamir uses probability vocabulary to state the chance of spinning a four-sided shape is greater than the chance of spinning a circle.

Instructional Focus: **Core** 1

Vanna

What is the probability of spinning a four-sided shape? Explain your answer.

4 out of 6 because 2 shapes do not have 4 sides

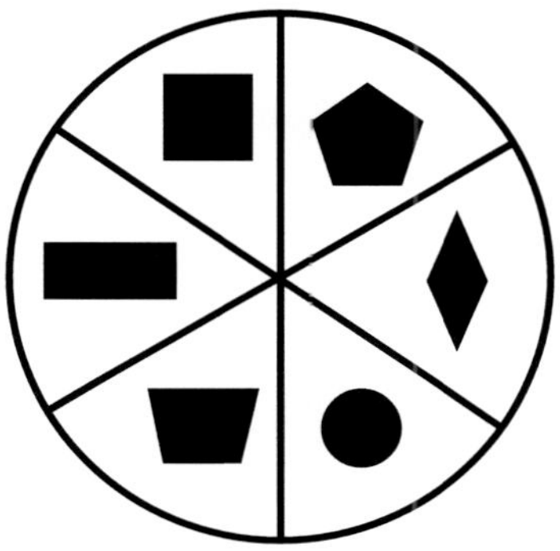

Vanna describes the probability using appropriate math vocabulary.

Instructional Focus: **Core 2**

Wade

What is the probability of spinning a four-sided shape? Explain your answer.

4/6 because 4 out of the 6 shapes have four sides. 4/6 is the same as 2/3.

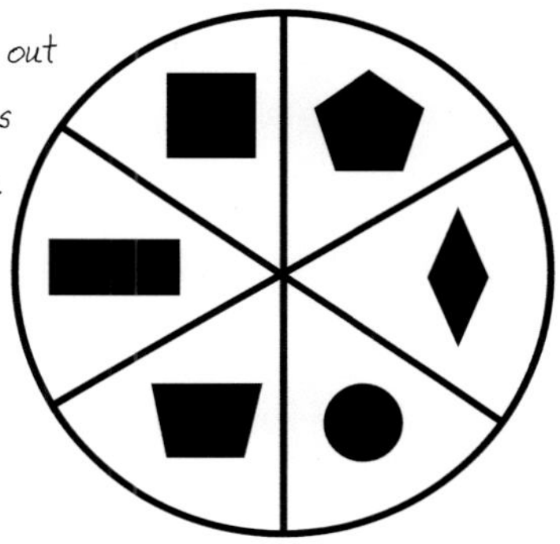

Wade records the probability as a fraction and simplifies the fraction in simplest terms.

Instructional Focus: **Challenge**

Data Analysis & Probability: Likelihood of an Event

	Rebuild Focus	Core 1 Focus	Core 2 Focus	Challenge Focus
	Mix & Match	**Blank Out of Blank**	**How Likely?**	**Spill the Beans**
Goal	Describe events as more, less, or equally likely	Describe the likelihood of events	Describe probability outcomes as fractions	Describe quantities as percentages and predict probabilities
Materials	♦ Cubes ♦ Cups ♦ Index cards	♦ Dice (variety of types)	♦ Bag ♦ 10 Cubes of 4 different colors (red, blue, green, yellow)	♦ Cup ♦ Bi-colored counters (red/yellow)
Suggested Activity Directions	Ask students to fill a cup with 10 cubes and describe the number of each color. Discuss all of the possible outcomes if one reaches into the cup and pulls out 1 cube. Ask students to decide which color is most likely and which color is least likely. Record the likelihood on an index cards and set aside with the cup. After making several sets of cups and cards, shuffle the cards and challenge students to match each card with the correct cup. Check each pairing and ask students if they match. Adjust as needed.	Provide time for students to explore dice with a variety of sides (e.g., hexahedron, decahedron, dodecahedron). Ask students to describe the likelihood of spinning specific numbers such as: ○ Even ○ <4 ○ >1 ○ 2-digit Encourage students to fill in the phrase _____ out of _____. For example, there is a 3 out of 6 chance of rolling an odd numeral. 	Choose 10 cubes to place in the bag. Ask students to consider all the possible outcomes if we reach into the bag and pull out 1 cube. Discuss with students how likely they are to pull out a red, blue, green, or yellow. Explain that the denominator represents the total number of cubes while the numerator represents the number of the specific color. Ask student to name the probability of each color as a fraction. 	Provide students with a cup and 10 bi-colored counters. Discuss percentages related to the number of counters. Ask students what percentage would be red if half the chips turned up red when spilled from the cup. Provide time for students to spill the counters and record the fraction and percent of each color. After several trials, ask students which combinations occurred most often and which occurred least often. 
Questions to Assess	♦ Which color is most likely in this cup? ♦ Are any colors equally likely within the cup?	♦ How likely is it to roll a value greater than ten? ♦ How many of the outcomes are even?	♦ What is the probability of getting blue? ♦ How would you decide what the numerator and denominator should be?	♦ What percent of the counters are red? ♦ Is it likely that 80% of the counters will be red?

7

monitoring
performance student progress

Conclusion

"If I only had more time!" This is likely the single greatest request among teachers reflecting on their instruction and the progress of their students. We constantly lament over the lack of time to teach, the lack of time for students to develop conceptual knowledge, and the lack of time to plan effectively for students. So how do we access this valuable commodity? The answer lies in the implementation of frequent and effective formation assessment. Formative assessment must occur on a regular basis to provide an immediate gauge of student understanding on specific objectives, and the format must be effective in gleaning specific information directing next steps for students.

Assessing and monitoring student performance is a complex task. The purpose of assessment is multifaceted. Formative assessments enable teachers to evaluate the effectiveness of the lesson, provide valuable insight into student learning, and offer a basis for future instruction. The constant administration of formative assessments and the analysis of student response informs a meaningful and targeted response to the class, to a group of students, or to individuals within the mathematics classroom. The use of formative assessment streamlines the entire instructional cycle. It delineates the difference between teaching a curriculum and teaching students. Teaching solely to the curriculum focuses on a specific standard and assumes all students enter a given mathematics class with the same experiences and background. However, we know that is not the case. Without formative assessment, we remain unaware of the gaps and misconceptions. Formative assessment allows rediagnosis and reteaching to be done in a timely manner. It is the benefit of meeting students where they are in order to help them progress toward a specific standard. This approach taps into that valuable resource of time because we prevent students from wasting time on things they already know or teaching too far beyond that point which enables them to bridge the gap.

Monitoring student performance through formative assessment enables us to provide a tailored instructional sequence altered to most appropriately facilitate learning. It is our hope that this resource provides examples of effective formative assessments, models of common student responses, and meaningful instructional responses to meet the needs of students. We hope this resource enables you to gain some valuable time by identifying and responding to student needs through formative assessment.

8

Appendix

Number Cards

0	1	2	3	4
5	6	7	8	9

Spinner

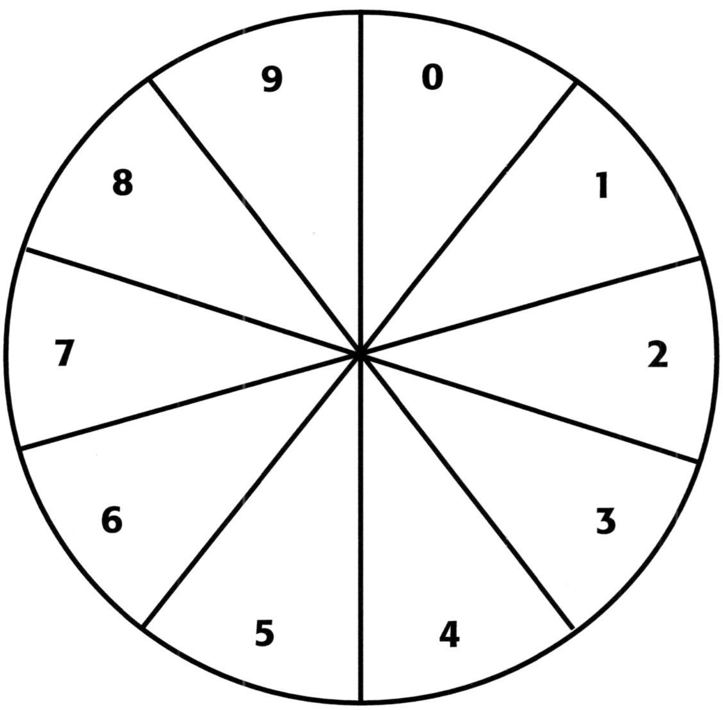

Operation Cards

All 24 students were asked to pay $7 for the field trip. What was the total cost of the field trip?

Sonya biked 18 miles in 3 hours. How many miles did she bike per hour?

Brad's family made 6 dozen cookies and shared them equally among 9 people. How many cookies did each person receive?

There were 837 beads in a total of 9 colors. If there were an equal number of each color, how many beads of each color were there?

The case of candies included 12 boxes of candy and each box held 15 candies. How many candies are there all together?

How could you use the calculator to find the product of 14x53 without using the multiplication key?

Pet Store Cards

Cut out the strips below and match them to the correct Pet Shop Card.

$\frac{1}{2}$ of the pets have only four legs.	$\frac{2}{3}$ of the pets are not spiders.
$\frac{1}{3}$ of the pets live in water.	There are $\frac{1}{2}$ as many fish as cats.
$\frac{1}{2}$ of the pets are cats.	$\frac{3}{4}$ of the pets have a tail.

Grid Paper

Multiplication Chart

X	1	2	3	4	5	6	7	8	9
1	1	2	3	4	5	6	7	8	9
2	2	4	6	8	10	12	14	16	18
3	3	6	9	12	15	18	21	24	27
4	4	8	12	16	20	24	28	32	36
5	5	10	15	20	25	30	35	40	45
6	6	12	18	24	30	36	42	48	54
7	7	14	21	28	35	42	49	56	63
8	8	16	24	32	40	48	56	64	72
9	9	18	27	36	45	54	63	72	81

Isometric Grid Paper

Target Game Board

Target Roll the die. All players must use the digit and place it in a space below. when all spaces are filled, calculate your score.	Score 0 points if false 1 point if true
____ . ____ > ____	
____ . ____ = ____ . ____	
. ____ ____ < . ____ ____	
____ . ____ ____ ____ + ____ . ____ < ____	
____ ____ . ____ - ____ . ____ ____ > ____ . ____	

Total ____________

If you played again, would you use the same strategy or a different strategy when deciding where to place each digit? Why?

Pattern Cards

Pattern Card #1

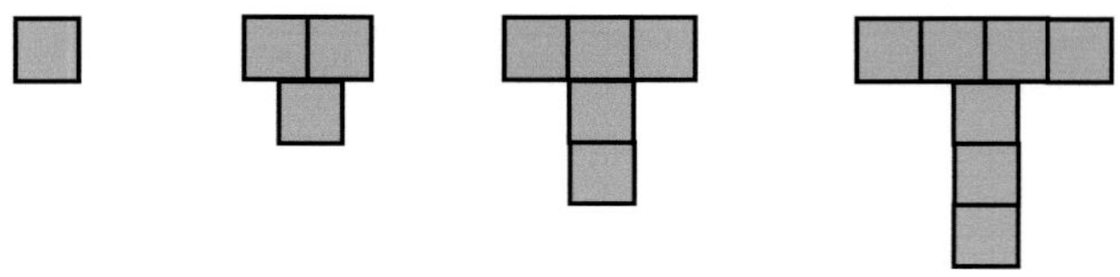

Pattern Card #2

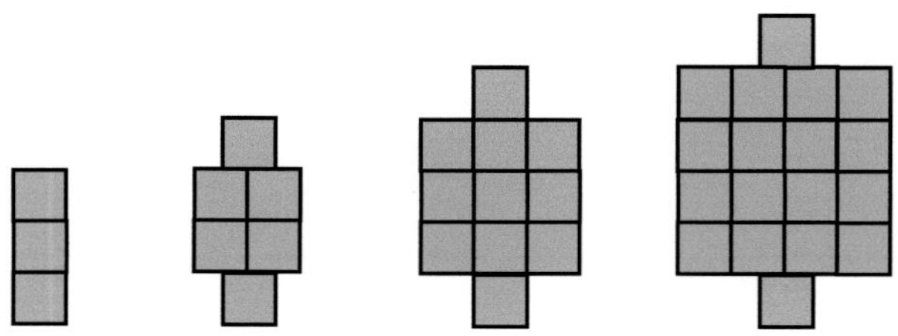

Pattern Card #3

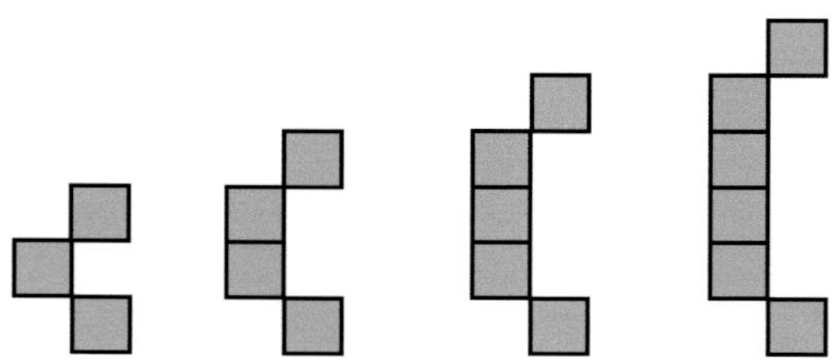

Pattern Card #4

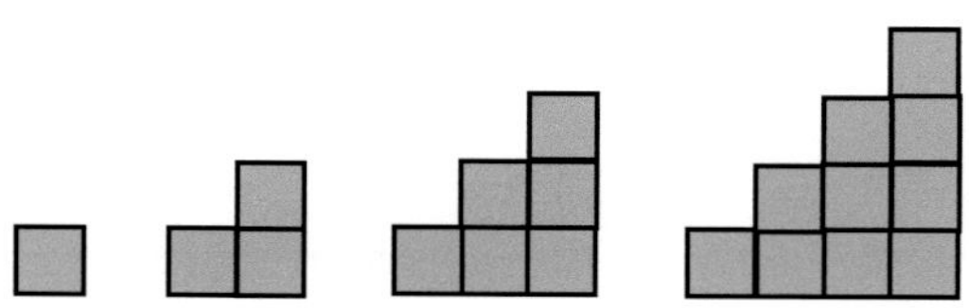

Dot Paper

Bibliography

Black, P., & Wiliam, D. (1998). Inside the black box: Raising standards through classroom assessment. *Phi Delta Kappa, 80*(2), 139–148.

Common Core State Standards Initiative (2010). http://www.corestandards.org/

Hattie, J., & Jaeger, R. (1998). Assessment and classroom learning: A deductive approach. *Assessment in Education, 5*(1), 111–122.

National Mathematics Advisory Panel. (2008). *Foundations for success: The final report of the National Mathematics Advisory Panel*. Washington, DC: U.S. Department of Education.

National Council of Teachers of Mathematics (NCTM). (2000). *Principles and standards of school mathematics*. Reston, VA: Author.

National Council of Teachers of Mathematics (NCTM). (2006). *Curriculum focal points for prekindergarten through grade 8 mathematics: A quest for coherence*. Reston, VA: Author.

Popham, W. J. (2008). *Transformative assessment*. Alexandria, VA: Association for Supervision and Curriculum Development.

Taylor-Cox, J. (2009). *Math intervention: Building number power with formative assessments, differentiation, and games*. Larchmont, NY: Eye On Education.

Taylor-Cox, J. (2005). *Family math night: Math standards in action*. Larchmont, NY: Eye On Education.

Notes

Notes

Notes

Notes

For Product Safety Concerns and Information please contact our EU representative GPSR@taylorandfrancis.com Taylor & Francis Verlag GmbH, Kaufingerstraße 24, 80331 München, Germany

Batch number: 10733489

Printed by Printforce, the Netherlands